Writing the Earth, Darkly

Ecocritical Theory and Practice

Series Editor: Douglas A. Vakoch, METI International, USA

Ecocritical Theory and Practice highlights innovative scholarship at the interface of literary/cultural studies and the environment, seeking to foster an ongoing dialogue between academics and environmental activists.

Recent Titles

Writing the Earth, Darkly: Globalization, Ecocriticism, and Desire, by Isabel Hoving
Dark Nature: Anti-Pastoral Essays in American Literature and Culture, edited by Richard J. Schneider
Thinking about Animals in the Age of the Anthropocene, edited by Morten Tønnessen, Kristin Armstrong Oma, and Silver Rattasepp
Romantic Ecocriticism: Origins and Legacies, edited by Dewey W. Hall
Linda Hogan and Contemporary Taiwanese Writers: an Ecocritical Study of Indigeneities and Environment, by Peter I-min Huang
Ecocritical Approaches to Literature in French, edited by Douglas Boudreau and Marnie Sullivan
The Green Thread: Dialogues with the Vegetal World, edited by Patícica Vieira, Monica Gagliano, and John Ryan
Interdisciplinary Essays on Environment and Culture: One Planet, One Humanity, and the Media, edited by Jean-Marie Kauth and Luigi Manca
Romantic Sustainability: Endurance and the Natural World, 1780–1830, edited by Ben P. Robertson
Ishimure Michiko's Writings in Ecocritical Perspective: Between Sea and Sky, edited by Bruce Allen and Yuki Masami
The Ecopolitics of Consumption: The Food Trade, edited by H. Louise Davis, Karyn Pilgrim, and Madhu Sinha

Writing the Earth, Darkly

Globalization, Ecocriticism, and Desire

Isabel Hoving

LEXINGTON BOOKS
Lanham • Boulder • New York • London

Published by Lexington Books
An imprint of The Rowman & Littlefield Publishing Group, Inc.
4501 Forbes Boulevard, Suite 200, Lanham, Maryland 20706
www.rowman.com

Unit A, Whitacre Mews, 26-34 Stannary Street, London SE11 4AB

British Library Cataloguing in Publication Information Available

Library of Congress Cataloging-in-Publication Data

Includes bibliographical references and index.
ISBN 978-1-4985-2675-3 (cloth : alk. paper)
ISBN 978-1-4985-2676-0 (Electronic)

∞™ The paper used in this publication meets the minimum requirements of American National Standard for Information Sciences Permanence of Paper for Printed Library Materials, ANSI/NISO Z39.48-1992.

Printed in the United States of America

Contents

Acknowledgments

This book has grown slowly, in response to the exhilarating developments in academic fields such as Caribbean studies, postcolonial studies, gender studies, and the environmental humanities, and I am deeply indebted to the many colleagues who guided my explorations in these fields. I would like to thank the Faculty of Humanities at Leiden University that funded a postdoc-project, my always inspiring, supportive colleagues and fellow travelers of the Department of Film and Literary Studies and the Leiden University Centre for the Arts in Society, and my coteacher Rob Zwijnenberg and the students who participated in the adventure of teaching and studying the environmental humanities. I am grateful to the many scholars who commented on earlier versions or presentations of parts of this book, offered me valuable advice, edited parts of the book in various stages, and/or discussed ideas; among them Mieke Bal, Elizabeth DeLoughrey, Maria Cristina Fumagalli and Bénédicte Ledent, Catherine Lord, Shani Mootoo, Nanne Timmer, Jonathan Todd, Ginette Verstraete, Gloria Wekker, Janet Wilson, the anonymous readers who commented on this manuscript, and earlier versions of separate chapters, and many others. My understanding of the environmental humanities has greatly benefited from the collaboration with my good friends at EASLCE and Ecozon@, and most recently the Benelux Association for the Study of Literature, Culture and the Environment, BASCE. I thank all those who edited earlier versions of parts of these chapters: Sean de Koekkoek, Jonathan Todd (again), Michael Katzberg, and Jonathan Smit. A big thank you, too, to Lindsey Porambo and Marilyn Ehm, the very patient editorial staff at Lexington Books, and Douglas Vakoch, for his support and encouragement. I owe special thanks to Olive Senior for granting me the permission to quote her wonderful poems. They were originally published by Calabash and Insomniac Press. Earlier versions of several chapters in this book were published elsewhere. A previous, shorter version of chapter 2, "Alterity," appeared as "Moving the Caribbean Landscape: The New Imagination of the Caribbean Environment in Caribbean Women's Writing," pages 154–68 in *Caribbean Literature and the Environment: Between Nature and Culture*, edited by Elizabeth DeLoughrey et al. (Charlottesville and London: University of Virginia Press, 2005). A much shorter version of chapter 7, "Trans," appeared as "Broadcasters and Butterflies: Sexual Transgression as Cultural Critique in Dutch Caribbean Writing," pages 477–513 in *The Cross-Dressed Caribbean: Sexual Politics after Binarism*, edited by Roberto Del Valle, Alcala Maria Cristina Fumagalli, and Benedicte Ledent (Charlottesville and London: University of Virginia Press, 2014). Both chapters are reproduced with permission of the University of Virginia Press. Fragments of chapter 4 first appeared in my "Remaining Where You Are: Kincaid and Glissant on Space and Knowledge," pages 125–40 in *Mobilizing Place, Placing Mobility: The Politics of Representation in a Globalized World*, edited by Ginette Verstraete and Tim Cresswell (Amsterdam: Rodopi, 2002). Chapter 8 appeared in an earlier, shorter version as "On Invasions, Weeds and Wilderness: The Dutch Imagination of Globalisation (Thrice)," 153–71 in *Global Fissures: Postcolonial Futures*, Cross/Cultures 85, edited by Clara A.B. Joseph and Janet Wilson (Amsterdam: Rodopi, 2006). Chapter 9 appeared in an earlier, shorter version as "Opacity and Openness: Creating New Senses of Dutchness," pages 297–312 in *Art and Visibility in Migratory Culture: Enacting Conflict and Resistance, Aesthetically*, edited by Mieke Bal and Miguel Á. Hernández-Navarro (Amsterdam: Rodopi, 2011). An earlier version of a fragment of chapter 10 was published as "Between Relation and the Bare Facts: The Migratory Imagination and Relationality," pages 179–90 in *Essays in Migratory Aesthetics: Cultural Practices between Migration and Art-Making*, edited by Sam Durrant, Catherine M. Lord (Amsterdam: Rodopi, 2007). These chapters and fragments are reproduced here with the permission of Brill Publishers. Excerpts from the collections *Talking of Trees*, published by Calabash Press, and *Gardening in the Tropics*, published by Insomniac Press and written by Olive Senior, are reprinted with permission from Olive Senior and the respective publishers.

Chapter One

Introduction

Writing the Earth in a Globalizing World

How to theorize the position of human beings in these times of intense globalization, climate change, and financial crisis? What discourses on identity and agency help us make sense of today's global predicament, and *why* are they helpful? For a long time, scholars approached postcolonial writing as if it were primarily concerned with working-through the ills wrought by human hands. But this reading fails to take into account the very many lively references to nature and the environment.[1] Postcolonial and/or Caribbean writers and artists often evoke the environment to grasp the newness of the globalizing world. Indeed, the writing of this book was inspired by the discovery of the ubiquitous, various, fascinating ways in which many contemporary Caribbean writers call forth images of nature to conceptualize globalization, and to rewrite violent discourses concerning race and sexuality. Olive Senior explores destructive global relations through the imagination of a violent garden, where one digs up human bones as easily as invasive plant roots. Shani Mootoo imagines hermaphroditic plants, insects and wild gardens to reimagine the nature of sexuality and race outside the dualist patterns imposed by colonialism. Jamaica Kincaid wrote a famous garden book to explore the ambivalences of colonial practices such as the traffic in humans and plants, and the colonial tendency of territorial control. Lawrence Scott evokes plants and flowers to analyze the entanglement of colonial desire and the homophobic suppression of desire. Michelle Cliff refers to gardens to explore marginalized sexual and racial identities, and the desire for a return. Aside from these relatively well-known writers, a range of lesser-known, Dutch-language Caribbean writers also evoke images of the (uncontrollable) Caribbean garden and the corrupted tropical environment as a means to express the ambivalences of exile and belonging. *Grande dame* of Surinamese literature, Astrid Roemer, grants metaphors taken from nature pride of place in one of the first self-consciously postcolonial masterpieces written in Dutch. She reads the postcolonial condition through the theme of (natural and social) corruption. In a short story, Ellen Ombre refers to flowers to offer a concise but complex image of global exploitative patterns. Cándani's novellas on the alienation of young Surinamese women in the Netherlands work through their main theme by evoking the sensuality of the Surinamese landscape and neglected garden plots. These texts are worthy of study because of the way in which they read post-, neocolonial, and global concerns through the lens of nature and an environment that succeeds so very well in capturing the dizzying complexity of the contemporary human condition. What is especially striking is the insistent presence of sexuality, either as a main theme or as an inevitable subtext. Indeed, all of these poetic explorations highlight the centrality of desire in colonial and global relations: the desire of being someone else or somewhere else, or the desire for the full control of the other. These desires are more often than not imagined as sexual desires.

The work of these, and many other Caribbean, African, and Asian writers suggest that a postcolonial approach alone is no longer sufficient. In a 2012 essay, Dipesh Chakrabarty offers his own reason for this: "the current conjuncture of globalization and global warming leaves us with the challenge of having to think of human agency over multiple and incommensurable scales at once." (1) According to Chakrabarty, the two earlier figures of the human (the Enlightenment view of man as a citizen entitled to rights; and the "postcolonial-postmodern" figure marked by differences of gender, race, class, etc., (2) no longer suffice to capture the figure that has emerged in this new era: humanity as a collective geological force. A force that has no ontological dimension, but is mere agency.[2]

Globalization (which some prefer to call neocolonialism) is not necessarily a completely new condition; it is often defined as a new stage in imperialism. Colonialism can be understood as that historical form of imperialism that emerged with the rise of capitalism, which necessitated a larger global field for its investments; its raw materials, labor forces, and the expansion of its markets (Loomba; McLeod). The transition to globalization heralded an increase of the destruction of vulnerable ecological and cultural communities all over the world that began during colonization. The global crisis is therefore not just economic and political, but also environmental, and not merely because of the climate change mentioned by Chakrabarty. The global exploitation of nature, the migration flows caused by large-scale environmental destruction, the intense changes and hybridization in local ecologies due to global trade, tourism, and global warming are all processes originating in, or exacerbated by globalization.

The insight into the environmental dimension has not been self-evident. Arjun Appadurai, for example, who called for a global imagination to replace the nationalist imagination of an earlier epoch, does not include the environment in the five "scapes" that indicate the "conditions under which current global flows occur" (1996, 37).

The importance of Chakrabarty's essay lies in its suggestion that the multidimensional processes of globalization call for a radical reconceptualization of the role of humans in the world, in which both social and environmental dimensions are taken into account. He was by no means the first to do so—for example, Ursula Heise's foundational study on the environmental dimensions of a "planetary consciousness" was published in 2008—but Chakrabarty's approach is especially helpful to understand how a large-scale ecological perspective radically innovates earlier forms of postcoloniality and globalization studies: the human subject of globalization can no longer be fully theorized through the postcolonial concepts of identity and difference, because they fail to address the huge *scale* of the new epoch. The environmental turn brings to the fore other key terms, namely agency and matter.

The writers I mentioned above move beyond the usual postcolonial concern for racial, gender, and sexual identity criticized by Chakrabarty, to ultimately question the fundamental dichotomy that sustains the others: that which exists between nature and culture. They do so, because the concept of the natural is traditionally called upon by authoritarian discourses of race, gender, and sexuality to legitimize their views. Caribbean writers will effectively address the dominant imagination of nature to criticize that legitimization—not only with the retort that race, gender, and sexuality would be social and cultural constructions, but with a reconsideration of what nature is, and of what is natural. Thus, they evoke nature as a queer, repulsive, seductive space of entanglement, where flowers, ink, blood, and words are all part of nature's corrupted but desiring flesh.

Traditionally, ecocriticism assumed that this alternative imagination of nature is aimed at overcoming the human alienation from *nature.* In contrast, postcolonial criticism primarily read such references to nature as a *cultural* critique. Both views simplify what is at stake. This writing should not be understood as a visionary blueprint for a new ethical relation to the global environment. It moves beyond the evocation of a new ethics, beyond discourse analysis, or the deconstruction of racist, sexist and homophobic discourses of nature and the natural. What makes many of these novels and poems so innovative and socially and environmentally relevant is that they reach out to a nondiscursive, elusive materiality, often presented as the zero-point of their writing. Many of these novels center around the protagonist's desire to return to the earth, to become matter, which is sometimes as much based in the death wish as in the desire to find a way of living that is not rooted in bankrupt, destructive discourses. More importantly, this desire also energizes the text itself. Thus, these texts suggest that, in times of globalization, it is only the dark, queer turn to *matter* that will allow us to find a way to exist.

Here, Caribbean writing comes close to the interest in matter and agency that characterizes the posthumanist approaches within ecocriticism. At the same time, however, this fascination with matter can also be read within the context of (post)colonial trauma, as I will show. In this study, writing is the point of departure, not theory; and writers will be inspired by a host of cultural and aesthetic influences that do not necessarily relate to the dominant theoretical discourses of the day. It is for this reason that this book stages a consistent interaction between Caribbean writing and criticism, postcolonial and globalization studies, ecocriticism, trauma studies, gender studies, queer studies, and transgender studies, in which none of these approaches has the last word. Each chapter will show how these approaches continue to interrogate and subvert each other. Thus, it is possible to trace the differences between the many roles played by nature and the environment in these innovative literary works, and explain why and how they help their readers to relate to the dizzying fullness of the global condition. For the sake of clarity, I will proceed by giving a general description of the three main ways in which the environment and nature are evoked: as counterdiscourse, as an interconnected web, and as matter.

Nature as (counter-)discourse

Caribbean literature has been studied within the context of colonialism and postcoloniality, within the context of a more regional approach, with respect to specific Caribbean epistemologies, and within the context of the theories of globalization and global interrelatedness. Recently, it has been rediscovered as a source of inspiration for ecocritics.

If ecocriticism has been traditionally concerned with the wish to find a new ethics that would heal "humankind's alienation from the natural world" (Buell, 1992, 8), my own reading of Caribbean literature has led me in a different direction.[3] In the novels I have analyzed, I found three different ways in which nature is evoked. In the first place, certain alternative images of, say, plant and animal life are evoked because they allow for different conceptualizations of race, gender, or sexuality. Here, the images of nature function as a cultural critique. In the second place, nature is sometimes pictured as a field of interconnectedness, marked by intense semiotic and communicative interaction, to which human beings can also relate. Thus, nonhuman organisms are endowed with subjectivity, as active agents within their own meaningful environment, and as potential partners in an interaction that need not be dominated by human intentions (see, in particular, chapter 2;

this approach may be intertwined with the first). This imagination can but need not be a proposal for an ethical way to live with nature. Thirdly, in the most dramatic cases, I found that nature was evoked to suggest an extra-discursive space that then functioned to subvert existing discourses—as sheer matter. At the same time, that space also functioned as the annihilation of discourse, or of the self. This evocation of space can well be read as a response to (postcolonial, global) trauma. That literary references to nature work in this third way, and that these references may be part of an effort of the literary text to overcome its own discursivity and to free up a space for a way to relate to globalization (without yet wanting to offer a new discursive understanding), is one of the most important points this book wants to convey.

The first function of references to nature I mentioned—that is, the cultural critique—will sound most familiar to readers with a background in postcolonial or cultural studies. They will recognize that there is a specific reason why nature, in the Caribbean, can function as the site for a critical counter-discourse. Caribbean literature and criticism is characterized by the imagination of landscape as a reservoir of repressed or erased histories (DeLoughrey et al., 2005, 1). The natural landscape, plants, trees, but also other material objects, testify to a reality (past or present) that was erased from dominant discourses. These histories cannot be recovered through scholarly archeology alone; as Alejo Carpentier, Wilson Harris, Antonio Benitez-Rojo, and others have argued, the imagination is needed to bring them to light. Sometimes, this imaginative creativity is understood as a spiritual visionary force. Some scholars refer to Vodun to suggest a spiritual reading of the natural environment.[4] One might see this approach to the environment as memory space as paradigmatic to Caribbean ecocriticism, as it was already developed in Caribbean/postcolonial theories.

Yet an astute reader will find that this is but one of the ways in which references to the environment can act as cultural critique. In the most critical novels, nature seems to take on an even more subversive role that exceeds its function as an indicator of a lost Caribbean history. In the cases I will discuss, nature is sketched as every bit as much creolized and corrupted as postcolonial culture, so that a social appeal to nature as the source for an ethics of authenticity, or a cultural identity, simply falls flat. And rightly so, for nature has never been pure and untainted. It has always been intertwined with the political, and vice versa. Writing about race *is* writing about nature; nature is often theorized as the ever-expanding network of varieties and diversities, of which some are identified and categorized as species or races. Since Linnaeus, nature is very much about sexuality too, as Londa Schiebinger (1993) explains. Until quite recently, we knew plants by their sexual organs. Studies of queer ecologies or queer environmentalism offer in-depth analyses of the intertwining of discourses of nature and discourses of sexuality. They have shown how, since Darwin, a healthy environment came to be seen as a space marked by intense heterosexual propagation; all forms of sensuality that fall outside that scope, are considered a deviant, undesired waste of energy (Gaard 1997; Mortimer-Sandilands and Erickson, 2005, 8–11). The exhilarating, momentous study by Bruce Bagemihl (1999), that brings together a baffling amount of research showing that the animal kingdom is characterized by a bewildering variety of sexual, affective, and parental behavior, and subsequent comparable studies (sometimes focusing on transgenderism in the animal realm) have so far not been able to change that dominant opinion (Roughgarden, 2004; Vasey and Sommer, 2006). This is one main reason why Caribbean writers address sexuality and gender when they wish to explore the question of how they can relate to the damaged Caribbean landscape, or how they can create a sense of home outside restricting notions of the natural. Images of transgender expression, bi- and homosexuality insert themselves in their imaginations of nature, whenever they write to uncover a noncolonial sense of self. Vice versa, when they want to address issues of race and sexuality, nature infuses their stories. Apparently, the search for a sense of the world that breaks with the heritage of a universalist, individualist, colonial modernity, leads to the radical reimagination of the most foundational issues: race, sexuality, and nature.[5] This imaginative exploration may include images, practices or expressions that could be considered non-normative, or violent, or repetitions of destructive colonial patterns. Whoever tries to comprehend all aspects of the contemporary imagination of nature in Caribbean writing therefore needs to seek support not only in postcolonial theory, Caribbean theory, theories of globalization, and ecocriticism, but also in theories of gender and sexuality, queer theory, transgender studies,[6] and, increasingly, in posthumanist thought. As Nicole Seymour (2013) argues, however, these approaches do not always go together comfortably (2–13). For the very reason that nature has been aligned with heterosexuality, queer theory, regrettably, "has worked against the development of queer ecological stances" (5). The close readings in this book support Seymour's well-argued conclusion that creative writers are very well able to work through these tensions.

This is only one of the ways in which this book will understand nature: as a discursive construction, and as the site for a potential counterdiscourse.

NATURE AS INTERCONNECTED WEB

Another way in which nature emerges in this writing comes closer to ecocritical priorities. It also answers the posthumanist interest in more than the human perception of nature, as it acknowledges that nonhuman organisms (plants, animals) are agents in their own right. Nature is now understood, not as space or matter, but as an interconnected web of intense sensual and semiotic interactions, in which human beings are, in principle, also implied. In a slightly different way, this ecocritical insight has lead postcolonial scholars to plead for an inclusion of ecocriticism within postcolonial studies. Canadian critic Susie O'Brien (2002) writes the following about the tasks of postcolonial criticism and ecocriticism:

> While postcolonialism [. . .] has done much to illuminate the ways in which all forms of representation cryptically articulate, as they attempt to transcend, underlying psycho-sexual and economic structures, ecocriticism speaks to the need to expand our understanding of those structures in order that we might address the broader ecological framework in which they operate. (181)

At the same time, ecocriticism must be well aware of the "tensions and contradictions that structure our engagements with the physical world" (181), examining these too, so as to avoid a simplified understanding of nature. The main point is the necessity of addressing the entanglement of the psychosexual, economic, and ecological. O'Brien points at different productive tensions between postcolonial criticism and ecocriticism, such as the following: where ecocriticism connects the local to the global in a simplified manner, postcolonial criticism helps to see the interruptions and distortions in those relations.

Both postcolonial criticism and ecocriticism embrace the world's materiality as a point of departure for analysis, but their understanding of that materiality is widely divergent (143). I will return to the latter point. British postcolonial scholar Graham Huggan (2004), who also argues for the inclusion of environmental destruction in the analysis of colonization and globalization, points out that postcolonial criticism helps uncover Western, liberal, and universalist tendencies in ecocriticism. Thus, postcolonial criticism represents a consistent critique of universalism within ecocriticism; whereas ecocriticism might offer a useful broadening of the concerns of postcolonialism, a different awareness of the world's materiality and a more complex view of the world's interconnectedness.[7]

If these scholars see postcolonial studies as an inspiration to enhance the critical force of the presumedly less politically inclined field of ecocriticism, others, in contrast, emphasize the radical potential in ecocriticism. Martinican writer, theorist, and activist Edouard Glissant (1990) insists that ecology offers the indispensable means to pronounce the formerly colonized's specificity (147), which he sees as an act of resistance. If colonization and globalization result in the erasure of local differences (and local complexities) for the benefit of a homogenizing transnational economic system, then ecology is the defense and reinstallment of local difference. The revaluation of the Caribbean natural surroundings, the protection of the land, and the defense of local Creole languages are three ways in which a colonial, composite culture like Martinique can defend itself against colonial violence that wants to erase that plurality (145). "[T]o oppose standardization [. . .] it is necessary to renew visions and aesthetics relating to the earth [. . .] this is where the imagination of an aesthetics of the earth—become indispensable" (148–49). Glissant admits that "a return to the earth" can also be anachronistic or naïve, reactionary or sterile (150).

Glissant's emphasis on locality is explicitly intertwined with this commitment to the basic interconnectedness of the world, which he refers to with the term "Relation." Glissant's concept of relationality, developed in response to Deleuze's theories, is a complex effort at conceptualizing the dynamics of globalization and creolization, without giving up the specificity of local places. Glissant's early evocations of relationality picture the cross-cultural world as a rhizomatic network of relations, in which the identity of every particular space, landscape, nation or individual is shaped by its cross-cultural and transnational relations to others, and is therefore always in the process of transformation. Thus, by imagining all locations in the world as linked, he also counters the fragmentation that marked the Caribbean after colonization. In good environmentalist fashion, Glissant explains the concept of Relation with the help of references to plants or trees. He juxtaposes the concept of identity that is called "identité-relation" (relational identity) with a more traditional, essentialist "identité-racine" (root-identity). The flexible, multiple mangroves, that grow entangled in each other, between air and water, would be a good example of the former, and can be opposed to the essentialist concept of identity that is symbolized by the fixed, seemingly independent European oak, with its deep roots. But while Relation is an immensely productive concept to grasp the interconnecting dynamics of globalization and the creolized Caribbean, critics have pointed to the fact that this interconnectedness, in later stages of Glissant's writing, develops into a universalizing movement that has lost its explanatory force. It leads away from the analysis of the specific agents involved in processes of globalization (Hallward, 2001), and as such one could say that the focus on connectedness hinders the view of specificity, and thus the world's diversity. Glissant's struggle to outline a "non-generalizing universal" illustrates the tensions in the Caribbean response to destructive postcolonial fragmentation, and the global interconnectedness that has both damaging and empowering qualities (Melas, 2001).

This struggle with the conceptualization of the world's interconnectedness is also at the heart of ecocriticism. Not all scholars are convinced of the productivity of some of the more influential concepts of interrelatedness. As Hallward's critique of Glissant's work argues, notions of connectedness do not necessarily have great analytical force. In addition, the romantic or spiritual longing for a harmonious reconnection with nature, which animates much traditional (predominantly U.S.) ecocriticism, is not very helpful when it comes to the analysis of the much less harmonious interconnectedness of globalization. Much Caribbean writing responds to the need to counter the destructive fragmentation of the region's community and environment during colonization, but—even though there are scholars who read this literature as if it were aiming at wholeness—there is also generally a keen eye cast at the paradoxes in the concepts of wholeness and connectedness (and this holds true for the early Glissant too). The Caribbean history of creolization and plurality forms a safeguard against simplification. I will show that this Caribbean distrust can even lead to the brisk statement that connectedness comes down to violence—and that the praise of connectedness is misguided and counterproductive. Rather than merely embracing the need for reconnection, then, Caribbean writing can be seen to explore the tensions between on the one hand the desires to unify and identify, and the desires to criticize and demolish coherence, unity, and identity on the other. This exploration often occurs through images of nature. The imagination of nature as connectedness is the second way in which the novels discussed in this book evoke nature.

NATURE AS MATTER

The most intriguing way in which Caribbean novels evoke nature is by aiming to call forth the earth's uncompromised *materiality*, both in response to trauma, and to create a space beyond the many confining and destructive colonial discourses that have defined it. In an earlier decade, ecocritics embraced literary evocations of nature because they resonated with their own desire for a *realist* connection with nature. They seemed to offer a way to steer free from the constructivist theories that approached the world primarily as discursive construction. This would have been the position of one of the U.S. founding fathers of ecocriticism, Lawrence Buell, who was targeted in a convincing critique by Dana Phillips in 1999.[8] Phillips's criticism was embraced by many ecocritics, especially those outside U.S. mainstream ecocriticism.[9] I endorse the critique of realism uttered by Phillips and many of his followers. The wish to escape constructivism may well come down to an unproductive resistance to theory. On the other hand, however, it has proved quite useful, and even downright necessary to question a merely discursive understanding of the world (a point made eloquently in Phillips's own work—for example, his work on waste).

Literary history itself is animated by such a recurring critique of discursive conventions. The evocation of nature's materiality can well be understood as an instance of a consistent literary desire to reach out to the "real." Perhaps we can even call this desire the very *raison d'être* of literature. That is, at least, what Viktor Shklovsky suggested in 1917, when he wrote about art's function to defamiliarize or estrange: "And art exists that one may recover the sensation of life; it exists to make one feel things, to make the stone stony. The purpose of art is to impart the sensation of things as they are perceived and not as they are known"[10] In a discussion about "things," however, literary scholar Peter Schwenger (2001) questions this opposition between perceiving and knowing: "Perception *is* a way of knowing," he argues (103), and even the most intense effort to experience the "real" is still shaped by one's social conditioning, and will not be able to establish an unmediated contact with that "real." Literary scholar Bill Brown (2001) supports his point. "The real, of course," writes Brown, "is not more phenomenal in physics than it is in psychoanalysis [. . .] beyond or beneath the phenomena we see and touch there lurks some other life and law of things, the swarm of electrons" (6). If we can be suddenly surprised by the physicality of the things and organisms around us (for example, when we bump into a tree), it is also true that we can only recognize these "things" through our complex social conditioning; it is society that tells us how to perceive and understand things (as lifeless objects, or as animated entities, Brown [9]).[11] Peter Schwenger's historical account of the wish to return to the "thing" brings us back to Heidegger's discussion of the manner in which things are "nominate[d] [. . .] *to* their being" (qtd. on 101) as meaningful objects. Things become meaningful through language, and in that naming, they inevitably lose their quality *as* thing. Nevertheless, the object still retains a "shadow" of its thingness, as Schwenger has it, in Heidegger's wake; we still sense "that there is an unknowable otherness to the thing" (103).[12]

Philosopher and cultural critic Slavoj Žižek goes a step further in explaining why the desire for the "real" is futile. In his reading of Lacan, he offers an enlightening definition of the psychoanalytical concept of the Real that makes it very clear that the desire for the "real" will never lead to stability: "far from being a substantial starting point and reference/guarantee, (the Real, IH) emerges as the retroactive effect of the failure of the symbolic process itself" (Žižek, 2001, 91). Thus, the Real would be that which threatens our symbolically constructed sense of self. It indicates the horrifying moments of collapse of our social order, and the self. In this sense, the Real highlights the irreparable gap between ourselves and the world. This psychoanalytical approach helps to understand why the literary evocation of the presence of the Real in literature can signify a crisis, and the collapse of identity. While the Real is connected to collapse and crisis, Žižek explains that, in spite of its anxiety-provoking and often horrifying nature, we need the Real as a support for the symbolic order (Žižek 1992, 31–32); we need the Real to answer, we need to read it as as sign. If the Real often threatens to traumatically break through the balanced Symbolic order, it is also that which supports that precarious balance (29).

For our discussion, the insight that the desire for the "real" often emerges from crises in the symbolic order, whether that crisis be individual, or collective, is crucial. While Žižek's Lacanian account helps to understand that the precarious balance of the symbolic order and the inevitability of moments of collapse is a structural aspect of human experience, and while the interest in the unknown or unrepresented is certainly a general characteristic of literature (see Caruth below), there are others who insist that the literary fascination for what lies beyond discourse is also motivated by historical, social events. That the yearning to return to the "real" is in itself already deeply historical can even be recognized in the academy. For example, it shaped different fields such as historicism, the studies of material culture and everyday life that emerged in the second half of the twentieth century—including, I would add, ecocriticism. In a 2005 essay, literary and cultural scholar James Berger explains why the desire for the "real" would emerge in times like these: it emerges in historical periods marked by rapid social changes, if not catastrophes, and what seems to be the collapse of the symbolic order (one of his examples is the violent event of 9/11). As Berger has it: "The damage to the symbolic order is not just structural [. . .] but is historical. Events have happened that have had lasting and symptomatic effects on modes of representation and communication" (345).[13]

Berger's essay offers a careful analysis of the role of (historical) trauma in this desire. He argues that one of the responses to "perceived damage to the symbolic order" may lead people to

> try to imagine modes of human life removed from symbolic-linguistic behavior altogether: in other words, modes of transcendence. [. . .] The transcendent, in this context, would be whatever surpasses, or passes below, all conceptual categories and symbolic forms. The divine, the sublime, the abject, the real, as well as the term *transcendent* itself, serve as catachresis for all that is beyond conceptualization but not necessarily beyond experience. (342–43)

His analysis explains why, in literature, "nature" (as that which exceeds discourse) can have the same function as the divine, the sublime, and the "real"; and why the environment, serving as the location of the sublime, and the life-changing "real," may be evoked as the privileged space beyond the broken symbolic order, that allows one to relate to a traumatic event. It is important to see that this literary perception of nature's materiality is principally different from the realist evocation of the environment as a place of healing. In the former case, the desire for the "real" is the result of trauma; and as the "language with which to summon and encounter" this otherness has lost its capacity to adequately represent, "the transcendent can only be expressed or addressed in terms of the traumatic" (Berger, 2005, 346).

The relation between nature (as the symbolic order's other) and trauma is at the heart of long-standing literary traditions, such as American women's writing that evokes the wilderness as the space where the unspeakable horror of women's lives is expressed (Blair, Knutson, Kolodny—see chapter 2). This link between nature and trauma is very explicit in the genre of the Gothic, but—more relevant to this discussion—it can also be observed in contemporary artistic or literary evocations of natural environments that once were the sites of, and witnesses to, atrocities related to the Holocaust (van Alphen), or colonization (DeLoughrey et al. 1).

The insight that trauma can most productively be analyzed as both structural and deeply historical is also at the heart of postcolonial trauma theory. In this sense, some scholars in postcolonial trauma studies criticize those trends in trauma theory that would rather focus on a "transgenerational, psychohistorical, timeless model of trauma" (Visser, 2011, 275), which would characterize dominant approaches, such as the work of Cathy Caruth, in which trauma is theorized as "a crisis of representation, of history and truth, and of narrative time" (Luckhurst, 2008, 5).[14] Some recent studies in postcolonial trauma studies suggest that a specific postcolonial approach would be in order. Here, trauma would be considered as historical and collective rather than individual. It would be more concerned with healing, through the verbal expression of the traumatic events (Visser refers to psychologist Judith Hermann) rather than with the unspeakability of trauma (as in Caruth, Lyotard; see Luckhurst, 4–13). It would grant more space to spirituality and ritual. In addition, as trauma would not be seen as unspeakable, there would be less interest in postmodern language experiments.

This study does not support such a crude differentiation between postcolonial and other narratives on trauma. In most novels discussed in this book, the traumatic events that motivate the expression of the unrepresentable materiality of the environment are rooted in colonization and (neocolonial) globalization. These novels relate traumatizing sexual abuse, homophobic exclusion, and incest to racism, (neo)colonial alienation, and involuntary migration. While the Caribbean novels and poetry discussed in this book are thus often concerned with the historical nature of traumatic events, there is no reason to discern an alternative postcolonial or Caribbean tradition that would approach trauma as *only* historically specific and collective, *only* within the context of healing, spirituality, and ritual, and *only* as in need of realistic expression. There is not just one way in which Caribbean literary texts respond to historical traumas; they use a multiplicity of textual strategies. Nevertheless, the shared interest in that which lies on the borders of the discursive (body, flesh, environment, matter) is striking.

The evocation of materiality that occurs in response to trauma should not just be understood as related to the unspeakability of trauma, however. It has special significance in the context of globalization, colonialization, and slavery. Decades ago, in the 1980s, postcolonial theorist Homi Bhabha already theorized the effects of colonization on the colonized's experience of space—private space is invaded and irretrievably broken open—and time—the split subjectivity of the schizoid colonial subject experiences time as halted and broken (Bhabha, 2004, 214–17). As I will show below, this schizophrenic colonial condition can lead to an intense, transcendent experience of matter (see chapter 5, Kincaid). In another vein, Caribbean scholars and writers such as Glissant, Gilroy, Wilson Harris, and Derek Walcott turn to the materiality of space and the environment as the location where meaning can be reforged, after acknowledging the traumatic nature of history itself (and specifically the traumatic nature of modernity, with the Middle Passage at its heart). In addition, decades after Fanon's famous theorizing of the traumatic anxiety caused by a white child's response to his black body (Fanon, 1952), the body itself has become a privileged focus of African and Caribbean literature, and of the literatures of postcoloniality and globalization.

Giorgio Agamben has famously theorized the state of the stateless refugee or asylum seeker as paradigmatic of the dynamics of globalization—*homo sacer*. In refugee and detention camps, human beings are reduced to bare bodies, vulnerable flesh. This is also the condition of existence during slavery, as Hortense Spillers explains (1987, 67). "[w]e [. . .] *were* capital," as Kincaid reminds her readers, just like sugar and cotton (1988, 37–38). Colonization, slavery, and the destructive dynamics of globalization may reduce people to flesh, matter. Below, in the section on "Earth," I will come back to this point.

The literary texts discussed in this book may effectively choose to work through these historically specific traumatic experiences by trying to touch base with the uncompromising violent realities they explore: with bare flesh, with the materiality of a land of which the history was destroyed. This literary strategy can take on different forms. It can be an exploration of the specific violence and obliteration involved. This need not take the form of a realist description. The traumatic events may often lead to a breakdown, and dissolution of the self in nature's materiality. In this case, the novel may evoke matter as the symbolic order's structural other. We can observe this especially in novels by Scott, Kincaid, and Cándani (chapters 3, 4, 5, and 7). Alternatively, it can also be a form of resistance. Facing and exposing the bare, material violence that lurks behind colonial discourses may shatter, or at least question, the very discourses that legitimize this violence. These strategies suggest that it is only through a literary confrontation with the incomprehensibly large, complex or shapeless, dark, queer materiality that the final break with the destructive *social* discourses can occur. These social discourses have lost some of their legitimacy, but not their violent force. Below, I will come back to the way in which the concept "earth" is used to capture this understanding of the materiality of the environment.

First, it is necessary to take into consideration two other possible theoretical approaches that are concerned with matter. These two approaches will also play a role in this book. The Marxist approach to materiality (the economic and political foundations of society) is apparent in some of the texts I discuss (notably Senior and Ombre, chapters 6 and 8), and it has always been prominent in Caribbean writing. It is an important dimension in feminist ecocriticism, too. In his 2012 volume on feminist ecocriticism, Douglas Vakoch, with reference to Buell, Heise and Thornber, offers a definition of feminist ecocriticism as a "politically engaged discourse" (Vakoch, 2012, 2). The prominence of scholars who integrate a critique of capitalist commodification and exploitation in their ecocritical analysis would support this view: Val Plumwood, Carolyn Merchant, and Greta Gaard, who are among the pioneers and founding mothers of the self-aware form of feminist ecocriticism that took shape in the 1980s and 1990s, all point at the destructive effects of (global) capitalism. Philosopher, scientist, environmental activist and feminist Vandana Shiva perhaps offers the most outspoken example of a politically committed form of global feminist environmentalism. In her 2010 overview of promising new directions in ecofeminism, Greta Gaard argues that there are nevertheless serious tensions between (metropolitan, predominantly white) ecofeminism and environmental justice. As the critique of global capitalism, and the related interest in economic materiality, is an intrinsic aspect of postcolonial criticism, which is one of the main approaches applied in this book, it will be a defining element of my close readings and interpretations.

There is another perspective that seems equally relevant to my argument: the posthumanist, new materialist approach that was articulated in the wake of Deleuze by scholars such as Elizabeth Grosz, Rosi Braidotti, Karen Barad, and Moira Gatens. This approach, which is part of a much broader material turn, can be considered as a new paradigm, with great promise for ecocriticism, as, for example, Stacy Alaimo and Susan Hekman have shown. Grosz helps us to differentiate new materialism from psychoanalysis and phenomenology. Phenomenology seems the more relevant approach for ecocriticism, and it has inspired important contributions to the field, such as David Abram's work. Psychoanalysis, in turn, can effectuate a useful disturbance of phenomenological assumptions. Psychoanalysis and phenomenology are, as Grosz (1994) argues, both concerned with the manner in which the body is experienced and given significance (*Volatile*, 116). There is, however, a crucial difference between the two affiliated approaches. Psychoanalysis, especially Lacanian psychoanalysis, opposes Merleau-Ponty's phenomenology because it "tries to restore the co-naturalness of man to his world, who centers on the corporeal presence," as Jacques-Alain Miller (2001) emphasizes. The main point of critique would be "that there is somewhere a place of unity, which is the identification of the being and body. [. . .] Psychoanalysis makes its space in the lack of this identification between being and body." We have seen that this is also an important issue in ecocriticism. Much ecocriticism is inspired by the possibility of a (phenomenological) reconnection of man and nature, mind and body. It hopes to heal the very alienation psychoanalysis sees as incurable. For, from a Freudian and especially Lacanian psychoanalytical perspective, alienation lies at the very base of human subjectivity, which is constituted through a series of castrations and alienating (mis)identifications.

New materialism offers a completely different perspective. Instead of asking how the body experiences itself and the world, this approach, which is part of a philosophical counter-tradition in Western philosophy, understands the body as a social object, as a series of relations that connects to other relations. That is, in Deleuze's words, as machine. The emphasis here is not on experience, but on action—this approach is interested in how the body *acts*, *connects* (Grosz, *Volatile*, 116). In this perspective, matter is no longer defined as the "real." This approach of matter as dynamic can already be found in the phenomenological writings of foundational theorists Merleau-Ponty and Bergson: "Active becoming is an emergent property of matter itself, its virtual development beyond its given properties" (Grosz, 2005, *Time*, 121). In new materialism, this approach is radicalized. In philosophy, in film and new media studies, in game studies, the environmental humanities and some regions of the digital humanities, we now witness the rapid expansion of a cluster of theoretical approaches that partake in the material turn and that all share a focus on agency, matter, and the posthuman.[15]

This vision of matter as becoming and the world as dynamically interconnecting is an important point of reference in this book. In spite of their productivity, however, these approaches raise a number of problems. In the first place, they should be brought into critical dialogue with scholars that insist on a reflection on their implicit political agendas, such as Slavoj Žižek, Peter Hallward, and Alexander Galloway. Hallward has shown the shortcomings of these approaches in a postcolonial context, as we saw when discussing Glissant: they might lead to an overly generalizing understanding of the world as connectedness that no longer has any explanatory force. Cultural theorist Alexander Galloway (2011) also urges us to reflect on the political agenda of some of the movements under discussion, such as speculative realism: "First, we need to acknowledge that this current round of realism is in fact a direct response to—and a desire to do away with—projects that were themselves extremely political, projects like identity politics. [. . .] So, I'm just nervous when explicitly political projects are kicked out, and what's left is a series of scientific claims to the real" (23).

In the second place, materialist approaches are not always appropriate. Indeed, some of the texts under scrutiny *are* very much about (Lacanian) lack, trauma, or experience, and I want to highlight these texts' specificity.

This book addresses the reasons why literary texts evoke the environment, and as I have shown above, references to the sheer, shocking materiality of the world play a role in the literary effort to come to terms with trauma. To understand this grappling with trauma, we have to take into account the psychological dimension, and psychoanalytical insights.

A reading that focuses on matter, agency, and the posthuman, when brought into dialogue with psychoanalysis, helps to bring out some of the most innovative (and disturbing) aspects in the Caribbean imagination of the human condition in the context of globalization.

WRITING THE EARTH, DARKLY

If we approach this writing as an effort to relate to the world's totality, or read its dark, queer dimensions as the wish to reconnect with matter, we will find that in either case this writing embodies the desire to reach out to the *earth*, rather than to the world—and this is the reason why the concept of earth appears in the title of this book.

The distinction between world and earth is central in Heidegger's work. If "world" is the concept that refers to the always historical, shifting context of meaning (or horizon of intelligibility) that allows us to experience our surroundings as significant, the concept of earth refers to the opacity that is *not* revealed in the world that is established by (for example) language (Heidegger, 1971; Karsten Harries, 1963). In his introduction to ecocriticism, Timothy Clark 2005) describes "earth" as "the very resistance to understanding and knowledge inherent to the non-human" (57). ["Earth"] "relates to *physis* at its most wild, resistant and opaque" (57).

The concept of earth receives new depth and relevance in postcolonial and Caribbean ecocriticism. In *Poétique de la Relation* (1990; *Poetics of Relation,* 1997), Edouard Glissant proposes "an aesthetics of the earth" ("poétique de la terre") to create an equilibrium between a focus on the local, and the inescapable openness to the global market, in the form of "international standardization [. . .] of consumer products" (149). In the introduction to their collection *Postcolonial Ecologies* (2011), Elizabeth DeLoughrey and George B. Handley evoke Glissant's concept of "aesthetics of the earth" in a way that helps to understand the two different dimensions in the concept that seem to me central to all postcolonial and/or Caribbean ecocritical approaches: the resistance of earth (as opaque matter) to discursivity, and the association with heterogeneity and alterity. First, DeLoughrey and Handley point toward the crucial place of opacity and corruption in Glissant's "aesthetics of the earth." Glissant opposes the "aesthetics of the earth" to reactionary and mystic celebrations of rootedness (1990, 147; DeLoughrey and Handley, 27), and to commodified forms of relating to the land. DeLoughrey and Handley explain the crucial difference of Glissant's approach by underlining that, for Glissant, the earth that must be loved is irretrievably destroyed by colonial exploitation; corrupted, urbanized, flooded (1990, 151), and ravaged (DeLoughrey and Handley, 27). I propose to acknowledge a close affinity with dark ecology's resistance to consumerist visions of the environment by sardonically emphasizing nature's shapelessness, abhorrence, and sliminess (Morton, 2007, 157–60; 196, 200; see also Senior's nature poems, Senior 1985, 2005). Glissant's suggestion that "surroundings" might be a more appropriate term than "environment" (1990, 151) testifies to his desire to speak in terms that resist facile appropriation of the environment as a coherent, well-shaped, balanced, stable whole. This, in turn, resonates with Heidegger's earlier understanding of the earth as undisclosed, opaque, resistant to signification (or commodification), even if the political context of his critique is very different.

In addition to this insight into the irreparable corrupted shapelessness of the earth, and building on it, DeLoughrey and Handley help to understand Glissant's "aesthetics of the earth" as a critique of totalizing discourses (28). Glissant's aesthetics of the earth, an "aesthetics of disruption and intrusion" (28), leads to the important postcolonial ecological insight that the world's radical *heterogeneity* should be brought out, rather than only its boundless interconnectivity. DeLoughrey and Handley usefully relate this critique to Spivak's theory of "planetarity" (2003, 77, qtd. 28), which also emphasizes alterity and heterogeneity.

In her important *Black on Earth: African American Ecoliterary Traditions* (2010), Kimberley N. Ruffin points toward a struggle with the concepts of nature, environment, and earth that is comparable to Glissant's. She highlights African American writer Haki Madhubuti's resistance to the terms "nature" and "the environment," and his decision to write instead about the relation between "human beings—specifically Black people in the United States—and our relationship to other people, the earth, and its resources" (qtd. in Ruffin, 6). His question is always: "Who owns the earth?" (qtd. in Ruffin, 7), and it seems significant that he uses the word "earth" instead of "world." For writers who want to oppose colonial, racial, and other exploitative discourses, the term "earth" refers to the material, nondiscursive sphere in a way that allows them to relate to it in a nonexploitative manner.

It is in this sense that the writers discussed in this book are more interested in addressing the *earth* than the *world*. In their work, the concept of "earth" as opaque, shapeless, and corrupt implies a radical critique of totalizing discourses and practices (colonization, globalization, commodification). We need to understand the term "earth" in its specific postcolonial sense, even if it still resonates with Heidegger's inquiry. "Earth," then, refers to a space of radical alterity that is associated with, on the one hand, the experience of being racially and colonially othered, and, on the other, with the inevitable resistance of matter to discursivity. Sometimes, as we saw above in our discussion of the relation with trauma, the earth even becomes the space of the collapse of the Symbolic. At the same time, the earth is conveyed in its planetary sense, as a global expansion that cannot be totalized. It is in this sense that this study intends to draw Heidegger's work, (dark and queer) ecocriticism, psychoanalysis, and postcolonial and Caribbean theories into a fruitful dialogue with each other, and with Caribbean creative writing.

"Earth"—in its sense of opaque shapelessness—represents a view of the environment that complements, disturbs, and opposes both the view that privileges connectedness, and the view that focuses on diversity. It will emerge in all chapters dedicated to either of the two views.

ARE WORDS AND NATURE NATURAL ENEMIES?

Apart from the interdisciplinary debates that run through all chapters of this study (the debate concerning the concept of connectedness, and the debate on matter and the "real") there are other important debates that have informed ecocriticism, which I want to touch upon briefly: posthumanism and language.

One of the most relevant aspects of new materialism, for many ecocritics, is the fact that it sides with posthumanism to render obsolete the divisions between language and sensuality, language and matter, human and machine, human and animal. These divisions have long been the subject of ecocritical debate. The inquiry into that basic duality between human/nature and human/animal leads to the consideration of posthumanism as a potentially useful approach for ecocriticism. Posthumanism is often associated with the technological developments that challenge the boundaries between human bodies and machines, and ponder the implications of the rapidly rising influence of IT, virtual reality, and the Internet. But posthumanism also addresses the boundaries between human and nature. A 2006 essay by Louise Westling opposes the two strands of posthumanism.[16] With an admirable eye for literary detail, she highlights the power of language to establish connections with different aspects of life. Her analysis invites a reconsideration of the duality between language and matter that new materialism also considers irrelevant. The divide between language and matter is already radically problematized, and even suspended, in Merleau-Ponty's phenomenology. Matter is no longer seen as something beyond language. Instead, language is seen as "the very voice of the things, the waves and the forests [. . .] language too is 'another flesh,' another 'wild being'" (Merleau-Ponty, qtd. Grosz, 1994, 102). But posthumanism and new materialism are not the only paths that lead to a less dualist perception of the linguistic and semiotic divide between the human sphere and nature.

Semiotician and zoologist Timo Maran, for example, states that it is impossible to differentiate culture from nature by arguing that the first would be characterized as a semiotic field, while the latter would exist outside the realm of semiotics and communication (which may be implied in the literary imagination of nature as nonhuman matter, as impenetrable "real"). An entire scholarly field (biosemiotics) "sees semiotic and communicative processes as an indispensable part of living nature" (459). Semiotic competence can be observed on the level of the cell, and it informs the communication between humans, plants, and animals. If we think of nature as interconnected, we should include the semiotic and communicative practices in this interconnectedness, varying from strategies of "biological mimicry and warning coloration" (460) to the nonverbal communication between a girl and her cat. Again, this brings us back to the problem of connectedness. In some (often early) ecocritical writing, nature's interconnectedness is praised as if it were "good"—as if writers should seek to express that ethical example. This ethical evaluation is less convincing when one considers that this interconnectedness often serves the violent goals of survival. As a plea for the recognition of nature's semiotic dynamics, however, it does make sense. Indeed, it would greatly benefit environmental protection if a plant's or animal's specific semiotic interaction with their ecosystem were to be respected. This respect is found in many novels included in this study; it is the basic precondition for the recurring imagination of the "return to nature." More immediately shocking, however, are the moments when nature is suddenly revealed as a nonsemiotic space, which happens in Kincaid's work. If one wants to understand the literary decision to picture nature as semiotic, or not, an ethical reading does not help. Literature is not primarily an ethical project, though it may be that too. But we will not understand literature if we do not approach it as *literature*. This study insists on taking literature's specificity seriously.

So, what *is* literature? Above, I referred to Merleau-Ponty's statement that language is flesh and "wild being." This intriguing statement might well be enriched with insights from the French poststructuralist inquiry into the libidinal investment in literary writing (Roland Barthes, Jean-François Lyotard). They lead to a different understanding of literature than can be found in much ecocritical work. If early U.S. ecocriticism was interested in realist representations of nature that would lead to a new ethics of the interconnectedness of man and nature, my own approach is less concerned with representation than with the performative and affective aspects of writing. I propose to read the manipulation of affect as central to literary writing, including environmental writing. This manipulation is a twofold strategy: on the one hand, it is aimed at an audience (for example, to subvert their habitual conceptualization of the natural); on the other hand, it can be understood as the writer's work with their own libidinal energies, to use the words of French philosopher Jean-François Lyotard. The latter dimension of the writing process, which often relates to the author's need to work through a certain anxiety or trauma, is not concerned with the critical response the text might wish to elicit on another level, from its readers. But this is what literature is too: for the writer, during the writing process itself, writing is a basically amoral manipulation of libidinal intensities. Ethical and critical aspects materialize at other moments of the process, that is, when a writer starts to contemplate their literary project, or to evaluate their writing, or when readers begin to read the text as a cultural artifact that interacts with other social discourses. Lyotard's concern with the way in which postmodern art relates to the unrepresentable (1992) helps to understand this essential dimension of literature. Trauma theorist Cathy Caruth defines the medium of literature itself in relation to the issue of representation. She argues that "literature, like psychoanalysis, is interested in the complex relation between knowing and not knowing" (1996, 3).[17]

From this perspective it is clear that ecocriticism cannot afford to only consider the representational aspects of references to nature, or their ethical and social effects. If we want to understand the full range of the many tasks that references to nature can accomplish in a literary text, we also need to take into account the psychological, phenomenological, and/or new materialist aspects of the *writing process* itself. It is less productive to study the degree of realism in a certain piece of nature writing than to analyze the nature of that particular desire to reach out to the "real." It then becomes possible to see, for example, that nature may be the site where the desire of a loss of self is articulated. Sometimes this is a regenerative loss of self (chapter 3). Sometimes, however, it is a return to a materiality that dismisses all social models of survival, a return that may be motivated by the death drive (chapters 3, 5, 6, and 10). These literary endeavors should not be mistaken for generalizable models for an interaction with the environment. They may very well be highly personal, untranslatable responses to a specific crisis.

From the perspective of poststructuralism (Deleuze, Lyotard) and new materialism, writing would be a practice that aims at connecting and disconnecting, in an endless exercise of the manipulation of narrative and libidinal energies, which are heightened or diminished by these moves. This will be an important principle throughout my readings. However, literature is, above all, a multilayered practice, and any literary reading should respect its complexity.[18] The layers of this practice may even contradict and subvert each other. This book, then, insists on exploring the varied psychological, ethical, political, spiritual, and sensual reasons why writers address nature and the environment—as healing interconnectedness, as a rich field for the rethinking of sexuality, gender, and race, as the unimaginably large scale of globalization, as the mind-baffling, traumatic, explosive, shapeless Real—and as life's zero degree, as death. These reasons may animate one and the same work, strengthening, confusing, or transforming each other. What this study hopes to bring to light, is that references to nature and the environment have become indispensable to any creative imagination that tries to deal with the scale and instability of the contemporary age of globalization. Nature and the environment are no longer just themes a writer may or may not address. They appear as the frame and the zero-ground of all writing and art that tries to reconnect with the complex materiality of globalization. Throughout the nine chapters of this book, I will unravel different aspects of the way in which literature evokes nature and the environment.

This book is organized in two parts, each dedicated to the exploration of a different concept that is crucial to discourses on the environment: connectedness and diversity, which I here present as both complementary and opposed. The debates on connectedness, matter, and language I discussed above run through all chapters.

The first section of the book is specifically devoted to the concept of connectedness. This concept receives particular relevance in the Caribbean, a region characterized by intense fragmentation and plurality, which are the result of a long history of violent colonial intervention. In the literary and artistic exploration of this fragmentation, and possible forms of reconnection, there is therefore more at stake than an aesthetic program. Artistic unity and coherence seem to respond to an aesthetic *and* a cultural and political need to heal colonial dualism and dividedness. They are produced in intimate interaction with political and social structures of dominance—based on discourses of race, ethnicity, class, gender, and sexuality.

The first chapter of this section, chapter 2, discusses the specificity of the environmental imagination in Caribbean-Canadian writer Shani Mootoo's *Cereus Blooms at Night* (1996). In its effort to lay bare and overcome the traumatizing racial and sexual dualism at the heart of colonial discourse, it offers an imaginative account of the interconnectedness of human and non-human life. If there is one novel in this book that tries to capture the intense sensuality of life within the sphere of insect, bird, and plant life, it is this subtle work. It pictures nature as *alterity*, which differs from the habit of imagining nature as "completely other" (which is common in romantic and pastoral writing). That dualistic move would call forth the problematic associations of nature with authenticity, wholeness, fullness, and so forth. The chapter moves beyond the mere statement that this imagination could serve as an alternative to colonial dualism. It traces the novel´s affinities with different environmentalist traditions, specifically those that deconstruct the notions of gender and sexuality. At the same time, the novel's emphasis on *indexicality* shows that it cannot be simply read as a *representation* of nature.

Chapters 3, 4, and 5 are dedicated to a more explicit critique of the concept of connectedness. Like Mootoo, the writers discussed in these chapters (Lawrence Scott, Jamaica Kincaid) explore the experience of postcolonial fragmentation and dividedness by bringing in the issue of sexuality, while intimately relating to concepts and images of nature and the natural. I will discuss their posthumanist approach by zooming in, ecocritically, on the place of nature in their writing: plants, flowers, fruits, birds—a nature that is inevitably understood as sexual. Instead of showing that nature is always also a cultural/political construction, these writers show the reverse: that cultural constructions—literature, stories—often imply imaginations of a nature that is sexualized. In addition, I argue that the writing process itself (that hazardous practice of connecting and disconnecting) is very much a libidinal venture. This leads to a different understanding of the text than a postcolonial, humanist reading that understands literature as primarily a practice of identity production.

One strategy of interconnection, which has been amply debated within postcolonial studies, and is prominent in nature too, is mimicry, a specific form of mimesis. Anthropologist Michael Taussig's famous analysis of the phenomenon of mimesis differentiates two complementary strategies within the phenomenon. According to him, mimesis refers not just to visuality (copy, mirroring—which is the traditional understanding) but also to corporeality (which would also refer, within the postcolonial and ecocritical context, to the much applauded strategy of establishing contact, and reconnecting). If *Cereus Blooms at Night* focuses on the corporeal aspects of mimesis, Lawrence Scott's novel *Aelred's Sin* (1998), against the grain, ponders the potential of the much more problematized *visual* form of mimesis as a means to reconnect the (post)colonial world. In chapter 3 I will answer Scott's call to look at the potentials of mirroring (as postcolonial strategy, see Homi Bhabha, or literary device) with fresh eyes, and discuss his exploration of the productivity of visual literary strategies as a means of addressing the postcolonial world, which is still divided by racism and (often religiously motivated) heterosexism and homophobia. Images of nature are central to his inquiry, which subtly unravels the manner in which a nonsubordinating, nonpatriarchal visuality may invite a sensual connecting between racially divided people. My reading highlights how Scott's evocation and intertwining of different Caribbean and European symbolist, modernist, intercultural, religious, and erotic discourses of *flowers*, allow him to evoke a sensual, embodied world, where the duality between religion and sexuality, visuality and tactility, subject and object, and blackness and whiteness is suspended. Ultimately, however, these visual and corporeal strategies of connecting do not lead to reconnection but to a loss of self. Scott's novel shows that this loss of self is at the very heart of both spirituality and sexuality. Nature, which is the site where spirituality and sexuality converge, forms both the place of the annihilation of the self and its possible regeneration. But Scott's story does not lead to the happy emergence of a healed self, but to death. As such, it exudes not only a warning to those who think too lightly about the possibilities to reconnect with nature, or to go against the racist, homophobic dualisms of dominant discourse; it also forms a powerful critique of those oppressive discourses, including the religious ones.

Chapter 4 considers the notion of connectedness as a response to postcolonial fragmentation by researching an attitude that is often seen as indispensable for the making of connections: the seemingly innocent concept of *empathy* (in a debate on Kincaid's *My Brother,* 1997). I will explain why the text relies on its many references to nature and sexuality to establish a connection with the Caribbean—and enter the critical debate on the question whether that reliance is productive or not. My main point is that Kincaid's texts show convincingly that identification and empathy will *not* effectuate an empowering reconnection with the Caribbean. Instead, Kincaid's work proposes to accept that others (including nature) can only be experienced from the outside. A reading of Kincaid's *My Garden (Book):* (1999) then brings to light her specific approach to the environment as lived space, that is, a space that cannot be known from the inside, but only by acting upon it. Kincaid's post-organicist view of the environment refuses the ideal of wholeness, but acknowledges the possibility of relating to it through action.

Chapter 5 continues that debate, this time by focusing on the desire, which is often expressed in Jamaica Kincaid's work but is made dramatically visible in the novel on her brother, to move outside discourse and reach out to the earth's materiality. For the narrator, reconnection with her Caribbean family seems only possible on the basic level of their shared mortality. During my discussion of Kincaid's narrative about her brother, I will show how Kincaid subverts the patterns of analogy and identification. Instead, her narrator found a way to relate to the intolerable opacity of her brother's destiny, first by distancing herself from him, by reducing and essentializing his identity in a politically problematic way, then by adopting a nonhumanist discourse, in which she understood his life in botanical terms. As she herself is a gardener too—an idiosyncratic, imperfect one, who is not in control of her garden—this move nevertheless allows her to relate to her brother's life and death. Not by any cultural or psychological identification, but because she, as a living being, shares her brother's materiality. My reading makes it impossible to praise Kincaid's work for its environmental ethics (or for its postcolonial merits in imagining a plural self—she is just not doing that!). Her writing shows, again, the pitfalls in the concept of connectedness, or in the endeavor to create a harmonious dialogue between humans and nature. For Kincaid, nature speaks strongest when it speaks of our shared, inevitable death.

The second part of this study is organized around a different concept, that can be opposed to the concept of connectedness that structured the first part: diversity. This second part will discuss the potential of approaches that aim at respecting differences and diversity, instead of emphasizing identity, and that aim at capturing the *specificity* of distinct places (theories of dwelling, transgender and transvestism, the concepts of diversity and opacity). More than the first part, this part acknowledges an important difference between mainstream ecocriticism and postcolonial ecocriticism: ecocritical scholars in the North tend to define their main problem as man's alienation from nature; ecocritical writers in the South propose another problematic: the destruction of the environment and the communities dependent upon it, poverty and oppression. Chapters 6 and 8 especially testify to that different approach.

I will also shift the debate to include a lesser-known literature: Dutch-language writing, notably by writers with a Caribbean background, and (for the sake of comparison) a few other immigrant writers. Why this selection of *Dutch*-language writing? My first reason is the wish to contribute to a more transnational, more *diverse* ecocriticism. Ecocriticism has long been a predominantly U.S. discipline. Postcolonial, Caribbean and other nonhegemonic American responses proved to be very productive in broadening and complicating this important field of inquiry. European ecocriticism, too, is now rapidly taking shape. European ecocritics testify to a different perception of the importance of nature in the imagination of the world than their U.S. colleagues (Goodbody and Rigby). For example, the fascination with the wilderness does not take the same place of pride. The heavily industrialized, densely inhabited Netherlands can be seen as an extreme example of the European environment: it is very much a constructed landscape, forged by centuries of close interaction between men and nature. How do writers respond to that environment, from which nature seems to have disappeared? What can ecocriticism be in such a place? How do immigrant writers from former Dutch colonies look upon the artificial, cold landscapes of the regions that were responsible for the destruction of *their* ecologies? What environmental aesthetics develop in such a context?

My second reason to include the Dutch case is the remarkable, and puzzling, celebration of openness in Dutch social discourses. This challenging peculiarity also makes Dutch-language writing a great case for the study of the diversity of environmentalist approaches. On the one hand, openness is praised as a quality of the Dutch environment, immortalized in the famous landscape paintings of the seventeenth century, with their wide open skies. On the other hand, openness and tolerance have become key terms in social and political discourses on immigration: the Netherlands present themselves, with modestly nationalist pride, as radically open, and ask immigrants to adopt the presumed virtues of openness, transparence, and tolerance. The rationale of postcolonial ecocriticism, in contrast, lies in the acknowledgment that all natural and social spaces are bafflingly complex, opaque knots of social, political, economical, environmental, and other, often global, interactions. How do we then understand this specific discourse? How do immigrant writers respond to this discourse, and where does the environment come in as an aspect of the debate on openness? The ninth chapter answers this question most directly, through a reading of some of the best-known hyphenated Dutch writers: Dutch-Caribbean writer Ellen Ombre, who is compared with Kader Abdolah and Hafid Bouazza.

Chapter 8 opens the debate on the ecocritical importance of the work of Ombre and Bouazza within an interdisciplinary discussion of the notion of diversity, which plays a central role in different discourses about globalization, notably in political discourse, environmentalist discourse (biodiversity), and the literature of migration. When Ombre's and Bouazza's texts are read within the large interdisciplinary frame of political and environmentalist discourses, they appear to react to basic assumptions about nature and the causes of diversity in both types of discourses. Instead of offering new imaginations of natural and cultural diversity, however, they explore the tensions between diversity and stability, nature and politics, and so forth, in remarkably different and innovative ways.

Chapter 7 looks at a specific figure that (as Marjorie Garber argued) often functions as a marker of social crisis, but that, as *Cereus* shows, can also *bridge* gaps, and thus overcome that crisis: it is a figure of reconnection. I will inquire into the role of the transvestite or transgender person (and other figures of sexual transgression) in the Surinamese and Surinamese-Dutch literature of the last two decades. This theme allows me to explore the extent to which different sexualities are naturalized or declared unnatural and erased, and thus the measure to which sexual diversity is expressed. My focus on this regionally and linguistically specific literature also enables me to trace the characteristics of its expression. This comparative approach becomes possible because the novel on which this chapter will finally focus, Cándani's *Huis van as* (*House of Ashes*) (2002) shares many affinities with two novels discussed in part one of this book, *Cereus* and *Aelred*. In these two novels, the transvestite, transgender, or homosexual person is closely associated with nature; in *Huis*, it is the sexually transgressive daughter-figure, involved in incest, who is placed on the side of paradisiacal nature itself. All three novels are about a deep alienation, caused by colonialism. The abundance of sensual references to nature is another stylistic similarity. The three novels are all involved in a radical deconstruction of alienating binary gender and racial oppositions through the lens of sexual transgression, but opting for different means to do so—*Cereus* by means of an exploration of the transgressive sexualities produced by or ruled out by colonialism, such as incest, and (in a different way) transgender expression and transvestism, *Aelred* by imagining the possibilities of homosexual love within a colonial religious discourse of intimacy and transcendence, *Huis* by reconsidering intimacy (as opposed to postcolonial separation and alienation), including its most extreme form: incest. What interests me most is the fact that, in *Huis* too, these literary practices of reconnection are tied in with a specific, radical reimagination of nature. To bring out *Huis*'s imagination of the Surinamese environment and the people who inhabit it, I compare and confront psychoanalytical, posthumanist, and spiritual readings. This view of the environment cannot be called a cultural critique; it offers a seemingly universalist ethical insight into the transience of life, and the existential tragedy of alienated, fragmented lives. Nevertheless, I argue that this view is shaped by a very specific imagination of a specific postcolonial condition.

Chapter 6 also clearly engages with the central theme of the second part of the book, diversity. It focuses on writers who are writing about their own local environment "from within." Through a comparison of the work of Jamaican-Canadian poet and novelist Olive Senior (especially her nature poetry) with that of others, notably Michelle Cliff from Jamaica and Astrid Roemer from Surinam, I will argue that a *postcolonial* "dwelling perspective" (Ingold) represents a very specific form of knowing postcoloniality and globalization, that neither joins postidentitarian mainstream theories nor remains steeped in colonial primitivism or nationalist or regionalist realism. In their writing, references to nature function as a means to convey and explore the concrete specificity of a space; not an essential authenticity, but a specificity that is understood as intensely historical and relational. These writers show a deep affinity with Glissant's evocation of images of nature as a means to ponder the opacity of place. Like him, they resist the essentialism that often goes with the celebration of nature as the sign of local authenticity. The natural environment, for them, is inevitably corrupted and contaminated. For Senior especially, this view leads to a pragmatic, culturally specific approach to what she understands, not as landscape, but as task space. Senior's work allows me to tie in with a larger debate on postcoloniality and ecocriticism. If the later Glissant, in the wake of Deleuze, used botanical metaphors to imagine the world's postcolonial and global interrelatedness, Peter Hallward pointed out that his visionary imagination of global dynamics testifies to a vision that leaves the postcolonial concern with the specific behind. Senior's poems about her efforts at laboring her torn, stubborn fields, full of human bones and hard knots, refuse the transcendence of that postcolonial writing that aims for the radically singular. Nevertheless, while she cannot be placed with posthumanism, Senior is interested in creating a non-anthropocentric universe, and she writes that universe into existence in her colorful poems, in which trees may speak and human bodies grow when buried. I would like to place her on a map of poetic positions, in the space where the critique of humanism and an approach of the (nonbinary) world as agency go happily together with a focus on the specific—as opposed to those positions (as in the later Glissant) where a critique of humanism and an emphasis on agency (becoming) is situated within a singularist frame (in Hallward's terms).

These nine chapters outline different traditions on the map of postcolonial and/or Caribbean environmental writing, or Caribbean and/or postcolonial writing that includes references to the nonhuman world. All of these texts, even the most politically committed, are written with an ear for what they cannot represent: the elusive materiality of globalization. This is the unarticulated space within which these works are written, and which decides both their desire to intervene in social discourses and imaginations, and their libidinal economy. If we want to understand the importance of ecocriticism as an intervention in other socially relevant forms of inquiry, we will want to consider the many different levels on which the desire to write the earth is at work.

NOTES

1. This book will discuss literary references to, mainly, nature, the environment, the earth, and mere "surrounding." See the introduction in Morton for an exploration of all definitions of nature. The term "environment" is used as the most general concept which links this inquiry with other cultural, ecological, and sociological forms of research. See below for a discussion of the concepts of earth and surrounding.

2. Dimock's 2006 study *Through Other Continents*, in which she pleads for an understanding of American literature within the framework of a much larger temporal and spatial scale than the usual modern, nationalist frame, has a great affinity with Chakrabarty's approach. I thank Ben De Bruyn, University of Leuven, for suggesting this work to me. Long before Dimock and Chakrabarty, in 1983, Caribbean writer and scholar Wilson Harris already offered his famous, self-consciously mytho-poetic proposal to read literature through its cosmic global connectedness (*The Womb of Space*).

3. Examples of ecocritical writing in this vein are manifold. One example can be found in Val Plumwoods ground-breaking work, which consistently addresses dualism such as nature/ratio. In one essay (2007), she presents the (otherwise seductively magic realist) argument that the main tasks for "a critical green ecological writing" (17) consists of "challenging [. . .] the sadodispassionate rationality of scientific reductionism" (17–18) and reclaiming and recovering "the space of everyday wonder and quotidian enchantment" (18) through a return to celebrating stones. She even states that "(t)he eviction of spirit and honour from stones and from the earth's body is one of the great crimes of modernity" (21). Though this statement sounds less extreme when read as an ecocritical critique of the duality spirit/matter, this definition of ecocriticism's task to restore enchantment is problematic. Restoring a sense of enchantment may be a poet's aim, but ecocriticism's task either lies in a rigorously interdisciplinary critique of the discourses and practices that led to global environmental destruction, and a reflection on the ways in which its academic practice is related to environmental action, or it is futile. Plumwood's own socially committed work is generally a good example of what a productive ecocritical approach can be.

4. See, for example, Knutson (1996) on Michelle Cliff.

5. Nicole Seymour's *Strange Natures* is a good example of an insightful study that departs from queer studies and environmental studies, but adopts an intersectional approach (including antiracism, but also anticapitalism, anticolonialism, etc.).

6. Susan Stryker and Aren Z. Aizura create a rhetorical link between ecocriticism and transgender studies in the introduction to the second Transgender Reader (2013): "Wildness, even more than cultivation or care, should characterize the health of our gender ecologies." (7).

7. Although Huggan's approach is not necessarily shared by most ecocritics, there is a wide acknowledgment of the productivity of a more intercultural approach, and of the need to include postcolonial writing in the field of ecocriticism, as Lawrence Buell shows in his 2005 study on the future of environmental criticism. During the last few years, many studies were published from an intercultural, nonmetropolitan, and/or postcolonial ecocritical perspective. Apart from Susie O'Brien, Huggan, and Huggan and Tiffin, I could mention Allewaert, Cami-

nero-Santangelo, Caminero-Santangelo and Myers, Cilano and DeLoughrey, Davis and Burke, DeLoughrey, DeLoughrey and Handley, DeLoughrey, Gosson and Handley, Gordon and Krech, Nixon, Roos and Hunt, Ruffin, and many others. As mentioned in note 6, Seymour insists on an even broader intersectional approach. Usefully, she also emphasizes the tensions between these approaches.

8. But see Buell, 2005, 32–44, in which he nuances his own position, and that of Phillips.

9. For example, the editors of a 2006 collection on European ecocriticism, Gersdorf and Mayer (2006).

10. "Art as Technique," translated by Lemon and Reis. I prefer this translation over the newer 1990 translation by Benjamin Sher.

11. Brown adds that in statements such as these, society takes the place of the stable ground for explanation. Indeed, any serious analysis should also ask *which* social aspect informs *which* aspects of such perceptions, in *what* historical and geographical context.

12. Schwenger then criticizes Heidegger's mystical emphasis on the synthesis that the thing effectuates, or embodies. Heidegger refers, writes Schwenger, to the "gathering or fusion of four elements: earth, sky, divinities, mortals" (107). To this interest in synthesis, he opposes Hegel's recognition of the multiplicity within the thing, the presence of "opposite aspects of truth" (107). This historical consideration shows that the desire to reach out to the thing can have a decidedly mystical dimension. This can even be observed in Shklovsky's work, perhaps due to the influence of philosopher Henri Bergson, whose thought also had some affinities with mysticism (severely criticized by phenomenologist Merleau-Ponty, who adopts Bergson's thoughts on many other points): "It is Bergson's apparent recourse to mysticism surrounding his understanding of intuition as a direct communion with things that repels Merleau-Ponty" (Grosz, 2005, 122).

13. He includes a reference to a useful statement by trauma theorist Dominick LaCapra, who also argues that it is important not to confuse (historical) loss with (structural) absence, so that it will remain possible to work through loss. This distinction is highly relevant for analyzing the response to trauma (Berger, 2005, 356; LaCapra, 2001, 45–47).

14. In the 2011 essay quoted here, Irene Visser takes issue with Cathy Caruth's presumedly ahistorical approach. However, in their introduction to the 2008 special issue of *Studies in the Novel* entitled *Postcolonial Trauma Novels*, to which Visser refers repeatedly, Craps and Buelens disagree, praising Caruth's work for its insistence that "listening to the trauma of another can contribute to cross-cultural solidarity" (2), and adding a quote from her foundational work: "*history* is precisely the way we are implicated in each other's traumas (*Unclaimed* 24)" (2, my emphasis). They do agree, however, that Caruth does not fully succeed in addressing historical and cultural difference: she does not follow up on her own plea for a cross-cultural approach.

15. In addition to Deleuze's philosophy of immanence or transcendental empiricism, and the new materialism discussed above, affiliated forms of science and technology studies and posthumanism (Donna Haraway, Katherine Hayles), and Bruno Latour's action-network theory, I could mention speculative realism (e.g., Quentin Meillassoux, Ray Brassier, Levi Bryant, Graham Harman) and object-oriented ontology (e.g., Graham Harman, Timothy Morton)—though the last two groups may prefer to see themselves as engaged in the speculative turn (Bryant, Srnicek and Harman 1).

16. At the same time, however, her essay reveals how hard it is to adopt posthumanist thought for ecocritical readings. Westling reads Virginia Woolf's modernist endeavor for transcendence, and the healing of the split human subject, as a posthumanist effort to unveil the basic interconnectedness of human, plant, and animal life. Woolf's work is then proposed as an example of the ecocritical imagination scholars in the field of ecocriticism should try to articulate. She does not elaborate to what extent Woolf's historically determined modernist project (with its emphasis on the healing transcendence of a painfully divided, alienated subjectivity) provides an answer to the equally historical postmodern, postcolonial epistemological crisis of our time. The postmodern, postcolonial insight that the world's fragmentation is irreparable (unless one embraces discursive strategies to create an illusionary coherence and wholeness), an insight that is also crucial to Donna Haraway's posthumanism, makes Woolf's historical answer less relevant—even if her diagnosis of that historical episode in modernity is still

admirably eloquent. Bergthaller's postmodernist exploration of anthropocentrism in the same volume *does* tune in with the postmodern.

17. See Luckhurst for a discussion of the different theoretical approaches to the relation between trauma and representation in literature.

18. Axel Goodbody (2006, 397) also underlines literature's concern with complexity, in an essay that testifies to the merits of literary analysis for ecocriticism.

I

Connectedness

Chapter Two

Alterity

Trauma, Sexuality, and Nature as the Unspeakable Other

Let us enter the debate on the imagination of nature in Caribbean writing by following a young trespasser into a central colonial trope: the Caribbean paradise.

> With its yard-long, bean-like purple pods, the mudra had taken over the side of the yard. [. . .] It took generations for a mudra tree to grow so large. The peekoplats hopped to the edges of the branches and their whistling subsided as though in curious and worried anticipation. [. . .] He suddenly felt himself a trespasser, an awkward voyeur. [. . .] It was as though he had stumbled unexpectedly on a lost jungle, and except for the odours he would have sworn he was in a paradise. (166–67)

The yard in the quote above, taken from the much praised 1996 novel *Cereus Blooms at Night* by Trinidadian-Canadian writer Shani Mootoo, recalls the image of the lost garden of Eden, which European explorers wanted to recognize in the lush Caribbean islands they visited for the first time. At the same time, however, we discern that there is something seriously wrong with this lost jungle paradise. It surrounds a house in a relatively ordinary village called Paradise, on the imaginary Caribbean island of Lantanacamara. Despite its unexceptional setting, this garden contains exceptional plant and animal life, life that is really only found in the heart of a primeval tropical rainforest. There is something very eerie about the garden, a feeling enhanced by the abhorrent smells that offend the visitor. In this ambiguous paradise, delicious odors mingle with the abhorrent stench of decay. The awed visitor is both repulsed and delighted. The ambiguity and otherness of this uncultivated garden is enhanced by presenting it as the counterpart to another garden in the novel, that is, the professionally maintained institutional garden in the yard of the almshouse, where much of the action takes place. The contrast between the two gardens emphasizes the eerie wildness of the jungle garden, the trees of which, we are told, "had sprung wherever birds and insects dropped their seeds" (124).

This ambiguous garden, which is at the very heart of Mootoo's novel *Cereus Blooms at Night*, has inspired several Caribbean and postcolonial critics. Sarah Phillips Casteel (2003) offers a lucid account of this "Gothic landscape" (19) as a threefold critique of the New World pastoral that refuses and subverts the colonial conflation of Caribbean landscape with a stable, essential Caribbean identity. In spite of its (also) Edenic nature, Casteel argues, this garden should not be mistaken for a site to reconnect to a lost, authentic self—it rather becomes the signifier of instability and uprootedness (25). If Casteel reads the novel within the framework of marvelous realism to understand it as a critique of the Edenic myth, Vera M. Kutzinski reads the novel as a magical realist to focus on the way in which it questions naturalness—notably by its Butlerian performative "'variable construction of identity,' where representations of different genders and sexualities intersect with the constructions of hybrid ethnicities that have come to characterize much of the contemporary literature of postcolonization" (174).

In Caribbean and/or postcolonial criticism, the Caribbean landscape is inevitably seen as broken and corrupted. These interpretations of the garden may serve as examples. However, the corrupted condition of this landscape is analyzed in different ways, and the proposals for an adequate imagination of Caribbean nature vary widely. The Caribbean feminist scholarship criticized by Kutzinski, for example, advocates a return to naturalness and authenticity (172), whereas Casteel's postcolonial ecocriticism and Kutzinski's queer postcolonial reading both refuse such an essentialist move, and propose a radical critique of notions of naturalness and essentialism instead.

I strongly endorse the latter position. The novel is critical to the point of suggesting that important insights cannot be *spoken*—that the ambiguity of life somehow exceeds discourse. The jungle yard, with its eerie plants, trees, and insects, does not function merely as anticolonial otherness but as (nonbinary) *alterity*.[1] However, would it be possible to move a step beyond readings that see the garden as a critique, to grasp the *specificity* of the alternative imagination it offers? Can we already differentiate between *different* forms of nonessentialism in the Caribbean imagination of nature, and describe these in positive terms, instead of as mere refusal? Surely, this specific garden does not only refuse colonial racist categories and exclusionary boundaries, but it also explores an alternative perception of nature, characterized by a *specific* sense of connectedness that we could tentatively try to situate. Indeed, gardens such as these have been read as an alternative, feminist experience of nature, in which connectedness and reciprocity are key terms. We can also discern an affinity with a queer literary tradition that offers the notion of a wounded landscape as a site to come to terms with life (Mortimer-Sandilands, 2005, 2). For there is reason to talk of a "gay pastoral" (Shuttleton, 2000), and it might even makes sense to talk about lesbian and gay practices of gardening (Cliff, 1980; Jarman, 1995 may come to mind). Even more convincingly, the novel has been read as the counterdiscursive exploration of an organic transgenderism that "bring[s] to the fore the natural dimensions of that process" (Seymour, 2013, 38), instead of siding transgenderism with postmodern technology, which is the dominant approach (42). This is possible because of the novel's project to "queer" nature, as I will explain.[2] Within the wealth of so many alternative traditions of practicing and imagining nature, can we trace the specific affinities of this particular imagination?

REPULSION AND DELIGHT: EVALUATING HYBRIDITY

A young man, one of the main protagonists of the novel, reacts with contradictory feelings when he enters the mysterious garden in Paradise for the first time. The garden is a smelly, tactile universe swarming with insects, reptiles, and unruly, creeping, rampant weeds. Remarkably, the most delightful sensual element in the yard—the cereus blossoms with their unearthly scent—is also the site of the yard's horror. The night-blooming cereus in itself is already an ambiguous plant: it is an invasive cactus, ugly in its bareness, and its blossoms seem to have been soaked in blood. But its smell evokes sensual bliss, and in this novel it will become the symbol of transgressive passion and indeterminacy. "A ghost of scent hovered close to the blossoms, remnants of last night's fullness now souring in the heat. Replacing the perfume was an even more startling one that seemed to emanate from the wall behind the blossoms. The smell [. . .] quickly grew nauseating" (165–66).

The story behind this repulsive smell is revealed when the reader learns about the cruel history of the dysfunctional family who once lived within these walls. As a boy, the father of the family, Chandin Ramchandin, was adopted by the British Reverend Thoroughly and his family, becoming a successful mimic man until his foster father refused to let him marry his daughter. Stating that the boy's love was in fact the incestuous love for a sister, the Reverend prevented their interracial love affair. The frustrated Chandin marries the daughter's (East) Indian girlfriend instead. This undesired wife, however, later leaves her husband to elope with the white daughter, whose lover she had become. Chandin begins to sexually abuse his daughter Mala, until, after many years, he dies during a violent quarrel with her. The daughter conceals his decaying body in the basement, shuts herself off from village life, and passes most of her years in the yard that grows around her father's corpse. The yard's intoxicating scents now intermingle with the stench of the father's corpse, which is in a state of incessant decay.

As the opening quote to this section suggests, Mootoo's yard exists outside of normative linear time. It is a place of decay and decomposition as much as of regeneration (123). Witness Mala's manipulation of time: "Every few days, a smell of decay permeated the house. It was the smell of time itself passing but lest she was overcome by it, Mala brewed an odour of her own design. [. . .] The aroma obliterated, reclaimed and gave the impression of reversing decay. [. . .] The odour hung, rejuvenating the air for days" (123). In resistance to decay, linear time is stopped. By reversing the decay before it reinstates itself, the natural cycle of endings and beginnings is interrupted and deferred.[3] In this strange timelessness, processes of growth and decay are extended infinitely. This explains why a body may rot for decades here without turning into a dry skeleton, and why the yard can harbor a mudra of such a size. Since the yard is thus turned into a space where nature takes on the form of a timeless primeval forest that exceeds the normal social order, one could indeed interpret it as the ideological, utopian image of an asocial feminine Nature. This image could then be understood as a woman's safe space, outside of the violent patriarchal sphere. That reading would tie in with the long history of the conflation of woman and nature, a history discussed by feminist ecologists such as Carolyn Merchant (1980). In her analysis of the landscape in Susan Warner's *Wide, Wide World* (1850), Andrea Blair (2002) refers to Julia Kristeva: Kristeva's concept of the semiotic, which is identified with the presymbolic maternal sphere, allows us to understand "the persistent use of the land-as-woman metaphor [as] a reflection of the subject's desire to reexperience the semiotic" (113).[4]

An argument in favor of this interpretation might be found in Mala's retreat from verbal and literary signifiers, until she "all but rid herself of words" (136). The sounds of nature touch her body directly, until "every fibre was sensitized in a way that words were unable to match or enhance" (136). At the very moment that the Law breaks into this wordless universe, the most traumatized part of Mala (personified by the thin girl she once was) tears itself free and literally flies away.

But the interpretation of this vision of nature as the desire for a woman's safe space misses the point. The novel is not structured around the binary opposition between pure, exalted nature versus soiled, perverted culture, and it does not oppose the yard's paradisiacal nature to the horrors of the house. As I have shown already, the yard itself is a place where life and death intermingle, where the abundant insect life, smells of rot and decay, and traces of dead animals combine to unsettle the squeamish visitor. The house itself is as full of spiders and insects as the yard, and the swarms of moths that are attracted by the cereus blossoms in the yard also cloak the incestuous father's body. Remarkably, the crawling insects have even found their way into the pages of the novel itself. The many sketches of insects in the novel seem as indispensable to understanding the narrative as the words themselves, and in this way the binary between the natural world and the cultural one is again subverted. Readers are warned that words alone cannot tell this story since it exceeds the boundaries of the cultural. But nature, on the other hand, cannot serve as healing fullness or authenticity either. Thus, the story takes its distance from the traditional ecocritical notion that the present-day global crisis is, as Helena Feder phrases it, the "result of alienation from our own senses and from nonhuman nature" (46). The crisis in Lantanacamara, which seems representative of the crisis in colonialism at large, does not stem from the alienation from nature and the natural, but from the violent definition of what is natural and whatnot, that is, the violent, colonial impositions of boundaries between the natural and the unnatural (as when the Reverend prohibits Chandin's love for his foster sister), and the violent crossing of these boundaries (as in the incest committed by Chandin).

Postcolonial theory, with its strong awareness of the irreparable ambivalence that came with colonialism, may form a remedy against the problematic ecocritical desire for a return to authenticity. Understandably, then, the narrative structure of *Cereus* will not offer its readers a comforting retreat from ambiguities. For example, the primary narrator, a male nurse who cares for the aged Mala Ramchandin, introduces himself as Tyler, "neither properly man nor woman but some in-between, unnamed thing" (76).[5] His endeavors to narrate Mala's story are disturbed by his desire to tell his own story, and as we learn about Mala's trauma, we also learn about Tyler's cross-dressing, recognized and encouraged by the old lady Ramchandin herself. Thus, the story is not just refusing a strict separation between nature/culture, but it is also narrated from the perspective of a border-crosser *par excellence.*

Instead of a naive ecocritical celebration of a utopian feminine natural space, as opposed to a patriarchal, perhaps colonial space,[6] the novel focuses on the topic of border-crossing itself; not on space, but on spatial *tactics*.[7] In this aspect, Mootoo's imagination could be related to other efforts by Caribbean writers to create a woman's space within hostile surroundings, such as Marlene Nourbese Philip's (1997) writings, and especially her brilliant and complex essays on what she terms "Dis Place." Heather Smyth (1999) shows the complexity of these efforts. Smyth concludes her analysis of Caribbean women writers in Canada by highlighting the insight (strongest in Dionne Brand) that a home can only be created by shared political activism. Philip's and Smyth's analyses suggest that such projects are complex and multidimensional, and should not be interpreted as merely the utopian imaginations of a woman's sphere. Thus, they too offer an argument against the interpretation of the yard as a woman's safe space, while supporting a reading that understands a woman's space as a practice, a difficult tactic.

The emphasis on agency, which is radicalized by new materialism, was already central to the kind of feminist ecology or ecological feminism advocated by Carolyn Merchant and Karen Warren. By redefining people as "relational and ecological selves" (Warren, 1994, 2), ecological feminism has attempted to overcome the notion that human beings (and especially men) somehow stand outside of nature. Merchant has emphasized reciprocity when she pleads for a partnership ethic in the interaction between humans and the nonhuman (*Earthcare*, 56), an approach echoed in Mootoo's work. Jim Cheney (1994), who formulates this reciprocal view of environmentalist ethics, helps to highlight the parallels between *Cereus* and ecofeminism: "For a genuinely contextualist ethic to include the land, the land must *speak* to us; we must stand in *relation* to it; it must *define* us, and we it" (174–75). On first sight, this is precisely what seems to happen in Paradise yard: "Mala's companions were the garden's birds, insects, snails and reptiles. She and they and the abundant foliage gossiped among themselves. She listened intently. . . . She did not intervene in nature's business" (137).

If this passage seems to express Merchant's and Cheney's ethics of reciprocity, Mala's decision *not* to intervene in the natural process suggests that the text is not really about a productive harmony between a woman and her nonhuman surroundings; the novel puts too much emphasis on the yard's excessiveness. Far from being a utopian representation of a reciprocal relation between woman and nature, Mootoo's novel is an example of a more general distrust among women writers of utopian perceptions of Caribbean landscape.[8] *Cereus,* together with some of the work by the Caribbean Canadian writers discussed by Smyth, problematizes the feminist "ethics of reciprocity" by offering a more wayward, painful notion of agency. The excessive space of the yard does not allow for peaceful interaction.

In the yard, both nature *and* the character Mala are represented as wholly other to the patriarchal, colonial, social, and symbolic order. Constructive reciprocity—that is, constructive border-crossing—seems an almost unattainable ideal in this space, as the excessiveness of this yard is an expression of the unspeakable excess of the border-crossing that victimized Mala. The strongest element in the representation of the yard in this novel seems to be its Gothic *intensity*. Because of this intensity, I would like to argue that the image of the yard is not only different from feminist utopias, but also more than an assessment of the creolized nature of the Caribbean landscape, or a critique of pastoral nostalgia, even as it signifies both.

One might suggest that this space represents more of a pathological retreat from traumatic reality, as Vera Kutzinski (2001) has done in her lovingly detailed essay on the novel. This would be in line with a long literary tradition that sees "the wilderness as a place of intimacy and rebirth" for

American women (Knutson, 1996, 284, referring to Annette Kolodny's 1975 study on the gendering of the American landscape). In a more negative vein, such a reading would also accord with the literary tradition that casts the landscape as the space where the unspeakable horror of women's lives is expressed (most visible in the Gothic). Andrea Blair (2002) thus reads the mountain in Warner's *Wide, Wide World* as the landscape in which the oppressed heroine "can express the inexpressible. [. . .] She can articulate what she is not allowed to in the valley. [. . .] In her anguish, Ellen becomes 'mad': that is, in what Kristeva describes as the 'body-to-body discourse with the mother,' the 'unsayable' language of madness is expressed" (120–21).[9] But if the yard is evoked as highly repulsive, seductive, unspeakable, and at the same time as the inescapable truth at the heart of a shared Caribbean reality, it is because the yard signifies the unspeakableness of the Caribbean condition in a more general sense. Though this great anxiety is a response to a personal trauma, it can also be understood as a response to the general Caribbean condition of ambiguity and hybridity, which forms the context of the individual trauma. To understand this anxiety, it is necessary to examine the ongoing debates about what can and cannot be considered natural. As Nancy Stepan (1982) and others have shown, these debates have already played a vital role in colonialism, particularly in the tropical environments of the Americas.

HYBRIDITY AND THE ORDER OF NATURE

Definitions of what can be termed natural have a special urgency when it comes to issues of sexuality and race. The sexual and the racial are intimately intertwined: Chandin is constructed as racially other in terms of the Reverend's refusal of marriage to his white daughter while his sexual desires are simultaneously labeled incestuous.

Cereus can be read as a radical intervention in these debates about the natural, especially those that uphold the essential purity and stability of a nature that opposes hybridity. As Nancy Stepan (1982) and Robert Young (1995) have explained, the ideologeme of racial purity has been central to Western debates on hybridity and race, making a tremendous impact on ideas (and legislation) about the social and cultural position of people of African, Asian, indigenous, or "mixed" descent. Young argues that colonial discourse constituted intertwining debates about race and sexuality. Nineteenth-century racial theories were derived from an emergent taxonomy of normative sexuality and naturalized desire. As Young explains, racial taxonomies arose out of a new science of colonial plant propagation and animal breeding. By 1840, European scholars began to defend the view that there were different human species and that the hybrid offspring of parents of different species would prove to be infertile, "a degradation of humanity [that] was rejected by nature" (Young, 1995, 15). In this example, nature is constructed as pure and inimical to hybridization. Sexual relations between people of different "races" were defined as contrary to nature's order.

When Chandin's white foster father exclaims that Chandin's desire for his white daughter is against the will of God (40), we can recognize this as a symptom of the anxiety with which Europeans often reacted to hybridization. Like many of his British compatriots, the Reverend rationalizes his anxiety in terms of a "natural" abhorrence of the unnatural, coded as a divine, God-willed order. Seen in the larger historical context, claims to the superiority of undiluted European culture justified colonialism and the slave trade, much as they constitute the Reverend's self-congratulatory missionary project in the Caribbean; "racial" intermingling would form a serious threat to this claim. The European obsession with Caribbean hybridity and degeneration can also be understood as a displacement of its fears regarding its own degradation and decline as a result of hybridization; these were processes Europeans thought to witness in other empires as Europe extended its colonial reach (Edmond, 2000, 42–43). These intense political and cultural fears were partly displaced through an obsession with health, sexuality, and ideologies of the natural. Thus the naturalization of "Nature" results in assumptions of purity, stability, boundedness—the opposites to hybridization.

Yet within environmental theories about "Nature," hybridization and instability appear as essential processes. Recent scholarship has shown that ecosystems are dynamic, unstable, and open systems, in which new species are incessantly settling, intermingling, and crossing with earlier species (e.g., Botkin, 1990). This approach has a counterpart in theories of culture. Caribbeanists have examined the age-old processes of creolization through which the plurality and diversity of the dynamic, open space of the region has been formed.[10] Decades after Edward Brathwaite's and Edouard Glissant's influential works, it is almost commonplace to consider Caribbean cultural identities as creolized. Nevertheless, David Watts's observation in 1987 that creolization did not serve as a sound foundation for "a regional cultural identity common to all inhabitants" (xviii) still applies in the first decades of the twenty-first century. For some it is a challenge to embrace and celebrate a creolized identity, either as an individual, or as a nation or region; and it is certainly hard to take pride in the hybrid character of the landscape. The influence of the colonial denigration of interracial marriage and hybridization perseveres.

Instead of arguing that an uneasy reaction to hybridization and creolization is misplaced, *Cereus* takes a closer look at the challenges involved in embracing hybridity. Similar to other Gothic novels, which evoke a European anxiety about the degradation of its culture, the text *deepens* the association between sexual and racial border-crossing and hybridity on the one hand, and madness and decay on the other. Simultaneously, the novel plays on the association between border-crossing and ambiguous beauty.

Responding to nineteenth-century racial theorists who held that degraded hybrids could be very beautiful but that their beauty was treacherous and unnatural, *Cereus* takes these two aspects as its narrative foundation. Through the image of the yard, the text explores precisely those sensations, repulsion and delight, instead of merely criticizing anxieties about racial and sexual border-crossing. Rather than celebrating nonviolent forms of border-crossing (such as the interracial love affair between two women), readers are invited to reflect on the nature of border-crossing itself.

To convey the ambiguity of this border-crossing, Mootoo's novel addresses sexuality head-on, because it lies at the heart of debates about Caribbean nationalism, nature, and hybridization. National discourses on the preferred forms of sexual relationships and family life are strongly prescriptive, taking recourse to traditional notions of the natural. But how does acknowledging the dynamism and creolization of all natural environments and the history of enforced hybridization change our understanding of border-crossing? Mootoo's novel raises questions about which forms of racial and sexual border-crossing will be acceptable and productive, and which forms will be seen as perverse and degrading. In what follows, I will show that *Cereus* proposes a very specific definition of both the natural and unnatural as an answer to these questions.

A QUEER WAY TO GO "BACK TO NATURE"

Ideas about natural sexuality have been shaped by a host of historical discourses, particularly natural history. Londa Schiebinger (1993) has pointed out that in the eighteenth century, Linnaeus's (hetero-)sexualization of plants, which became the basis of his botanical taxonomy, "came to recapitulate the sexual hierarchy of Western Europe"(4). If one turns to nature for models of a natural sexuality, one will often find oneself peering into a reflection of *social* ideas about sexuality (see also Mortimer-Sandilands, 2008, 457 for the turn to nature to find support for social ideas). The reconstruction of nature as a racial, sexual, and gendered hierarchy was motivated by the desire to find an authorization of social ideas: "Nature and its laws spoke loudly in this age of Enlightenment, where philosophers attempted to set social convention on a natural basis. [. . .] Scientists [. . .] took up the task of uncovering differences imagined as natural to bodies and hence foundational to societies based on natural law" (9). Eighteenth-century botanists "gave undue primacy to sexual reproduction and heterosexuality" (22), downplaying the fact that most plants are hermaphroditic, and not "male" or "female."

Catriona Mortimer-Sandilands (2005) explains how the rise of evolutionary thought in the second half of the nineteenth century lead to the conviction that heterosexuality is the central criterium for a healthy natural environment. "The logic goes like this: If the ability of a species to survive in its environment is tied to its reproductive fitness, then 'healthy' environments are those in which such heterosexuality flourishes" (7). In her 1997 essay on the need for a queer ecofeminism, Greta Gaard also argues that "'natural' is invariably associated with 'procreative'" (120). This assumption still determines dominant opinions of healthy and natural sexual behavior, and healthy and natural environments—in spite of the immense amount of research that has demonstrated the mind-boggling diversity of sexual behavior in nature (Bagemihl, 1999; Roughgarden, 2004).

Theories about nature and what is natural are integral to colonial discourse, where they are employed to build authoritative arguments about the organization of society. Toward the end of the nineteenth century, "(b)iological theories of decline were becoming the dominant form of social critique" (Edmond, 2000, 44), and worries about social and political stability were expressed as an anxiety about the condition of the "national body" (45). But we can also note opposing strategies: arguments based on observations of nature can be used to create a critical argument against present-day dominant discourses about sexuality and race.

Taking recourse to a convention that was also used by the modernists,[11] Mootoo questions these naturalist discourses on their own terms. Instead of referring to nature in order to picture the degradation and decline of hybrid cultures and persons, *Cereus* can be read as a plea to acknowledge the inevitability of sexual and racial border-crossing and the resulting intermediateness and hybridity, however ambiguous and uncanny they may seem. It is also a plea to differentiate between the ambiguous and the violently perverse. To make its point, the novel focuses on a specific flower, that sexual part of the plant favored by eighteenth-century botanists to make *their* point about natural sexuality.[12] To create a different discourse about sexuality and hybridity, the novel takes the night-blooming cereus as its central image—an ugly cactus with beautiful flowers that only bloom during one or two midsummer nights each year, and that exude an overwhelming scent.

The cereus, like all cactus plants, is a hermaphrodite; it can boast of both a stamen and an ovary, parts usually designated as respectively male and female. In the novel, it also becomes the central token of border-crossing.[13] The first cereus clipping we hear about is brought to the aged Mala Ramchandin in the almshouse by her visitors Otoh Mohanty and his father Ambrose. Father and son form a curious pair, and they embody the novel's most remarkable instances of border-crossing and mirroring. Ambrose is a former suitor of Mala, but he proved unable to save her from trauma, and he has slept away most of his life ever since. Otoh is his remarkable son: a transgender boy, who (almost) succeeded in letting even his parents forget that he was assigned another sex at birth. Otoh does not just cross the border between genders, but also between generations. He takes over his father's role by bringing Mala food, sometimes even dressed like his father, in an effort to make amends for his father's betrayal. The cereus is therefore a gift from two people who cross borders between genders and generations, past and present, sleep and waking life. Earlier, the cereus was used as a gift from the female lover of Mala's mother, who wished to delight her and her sister. The cereus clipping plays a role in a significant exchange between the almshouse gardener and Tyler, who begin to reach a mutual understanding when the gardener tells Tyler about his gay brother. The cereus both covers and marks the incestuous father's body behind the wall against which it has grown, linking the decay and delight of Mala's space (the yard) to the unspeakable horror of the father's space. Finally, the cereus is the plant that regulates the blossoming love affair between the transgender boy Otoh and the cross-dresser Tyler—they promise each other to hold off the consummation of their love until the cereus will bloom. In this way, the cereus marks and mediates many transgressive relations, linking peoples and spaces that are not considered naturally related. Apart from its great ambiguity (as both excessively repulsive and delightful), the cereus itself is not a hybrid plant. Without being a hybrid or an "alien" border-crosser itself, as it is native to the West Indies, the cereus still functions as the inspiration for the novel's many border-crossers. In Peircean semiotic terms: the cereus is not an *icon* of the hybrid, but it is an *index* of border-crossing (as it is all but touching the evidence of that transgression—the father's corpse). As an ambiguous night-flower, it is a *symbol* of the hidden and the transgressive.

The fact that the cereus is indigenous to this part of the world suggests that ambiguousness is at the heart of the Caribbean. And, indeed, Mootoo's Caribbean world abounds with queer, hybrid, border-crossers *par excellence*. The novel is a dizzying game of endlessly crossing sexual boundaries, boundaries of gender, and even the boundary between human and nonhuman life, where characters are likened to plants or animals. As a result, ambiguities and mistaken identities abound. A man (Chandin Ramchandin) "mistakes" his daughter for his wife. A mother (Chandin's wife) "mistakes" a woman from another race for her husband. The parents of the transgender boy "mistake" their gender son. The son is mistaken for his father. There is the transvestite (or transgender) male nurse. Finally, there is the abused daughter herself: Mala takes on birdlike qualities, and at times even seems to become one with a tree. This is not all, however. For these characters also identify with each other: Tyler, the nurse, finds in Mala Ramchandin a "shared queerness" (51), and Otoh, too, states: "I felt as though she and I had things in common. She had secrets and I had secrets" (133). Indeed, these three characters, recognizing in each other different forms of in-betweenness, different kinds of becoming ("becoming-animal" and "becoming-plant", "becoming-woman" and "becoming-man"),[14] are in comparable positions, though they are not equally traumatized. But the novel suggests, through the words of Otoh's mother, that many more citizens may share that condition. She addresses Otoh: "You grow up here and you don't realize almost everybody in this place wish they could be somebody or something else? That is the story of life here in Lantanacamara" (258).[15, 16]

In other words, in Mootoo's Caribbean, border-crossings are inevitable. Natural and social life are represented as an interconnected web in which intimate contact constantly leads to disguises, transformations, hybridizations and border-crossings. Since many of these borders were imposed by colonial administrations and ideologies, Mootoo's novel creates a different context for the evaluation of these sexual and other border-crossings. She represents a return to nature—but an unexpected, unconventional, queer return, which brings with it a completely different notion of the natural from colonial or nationalist views. In *Cereus*, sexuality takes on diverse forms (between people of different "races," of the same sex, between a boy-who-has-been-assigned-another-sex and a cross-dresser, between night-blooming cereus and the night-flying insects that pollinate it), that it becomes impossible to point out one normative sexuality. These diverse expressions of sexuality all appear to be in agreement with what happens in nature. The novel suggests that the only perverse forms of border-crossing are produced by the violence of colonial sexual and racial laws. Consequently, the refusal of hybridity, and thus a denial of exogamy, as in the Reverend's refusal to let the East-Indian Chandin marry his white British daughter, leads to the ultimate form of endogamy: incest.[17] This is a cynical comment on the violent consequences of the colonial fear of border-crossing.

Cereus shows how the painful hybridity and ambiguity of nature cannot be healed, for they are the result of the history in which the Caribbean has been violently interpolated. Instead of trying to return to an organicist and harmonious sense of nature, *Cereus* works through the pain caused by colonial destruction and transformation, subtly differentiating the queer from the violence of the perverse. Its exploration ends up suggesting a productive way to cross borders: with great care, and always open to the reactions of others involved. We might be reminded of a comparable fragility in Mortimer-Sandilands's reading of Jane Rule's queer writing "about wounded people coming to 'belong' in a rural nature by [. . .] embracing *its and their fragility*, and by thus mindfully and daily recognizing the contingencies and tensions of their web of attachments, including their implications on other-than-human communities" (2008, 464, emphasis in original). The acknowledgment of the fragility of this "web of attachments" strikes me. In the same vein, the two border-crossing heroes of *Cereus* can only become lovers by means of a tentative, mindful, interactive transgression. In the midst of that process, the novel suggests that a happy end could be possible—though it is not a return to authenticity and purity. It is here that *Cereus* shows its modest, critical utopian dimension.

This analysis shows that we can indeed move beyond the statement that *Cereus* is a critique—though it is indeed a highly intelligent, seductively sensual, provocative critique of tenacious colonial binaries, that is, a courageous *working through* of that still influential and destructive heritage, that shapes the postcolonial, globalizing world. We can begin to recognize that *Cereus* offers an imagination of the human and natural world as intensely connected—though it is clear that this description is far too general and coarse to do justice to the novel's subtlety. There are many wildly divergent notions of connectedness, from those in nostalgic pastoral traditions to those in ecofeminist thought. If I were to sum up the approach to nature most clearly articulated in Mootoo's novel, in ecological terms, I would argue that nature is not presented as an essence from which one could deduce norms about what is natural, but as transgressive interrelatedness; an unstable, inharmonious, damaging and damaged, productive, vital, queerly erotic interrelatedness, in which impurity is the norm.[18] This nature can never be fully described or appropriated. Mala's garden is destroyed when scientists and traders enter it to capture, categorize, and sell its riches. In this novel, nature is ultimately presented as society's ungraspable, nondiscursive, subverting *alterity*.

A last question remains. I have read the novel as a critique of certain social (racist, sexist, heterosexist) ideologies of naturalization—more as a cultural critic than as an ecocritic. But what would be the environmentalist relevance of this imagination of nature? Does *Cereus*'s literary imagination of nature help readers come to terms with the environment that has been damaged by globalization? Could we relate this poetic representation of nature to environmental discourses on the Caribbean? And if so, what is the role of literature in debates about the environment?

CONCLUSION: NEW PLOTS FOR LIVING WITH ENVIRONMENTAL CHAOS?

Like Shani Mootoo, feminist ecologist Carolyn Merchant is driven by the wish to develop an intimate interaction with the unpredictably interrelated world. In her eyes, chaos theory has put an end to a simplistic linear understanding of nature. It "fundamentally destabilizes the very concept of nature as a standard or referent. It disrupts the idea of the 'balance of nature,' of nature as resilient actor or mother who will repair the errors of human actors and continue as fecund garden" (1995, 54). As in Mootoo's novel, there can be no question of recovery. A "recovery-plot" is unthinkable, a happy return to authenticity is impossible. What other plot might then be appropriate? "What would a chaotic, nonlinear, nongendered history with a different plot look like?" Merchant asks (55). She muses that such a plot can "only be acted and lived, not written at all" (55). But does that mean that there is no use for literature—that source of new, written plots—within environmental debate?

Perhaps there is. Merchant's ethical reflection on the way to relate to a chaotic, interrelated world ties in, for example, with Edouard Glissant's (1990) poetics of Relation. The importance of Glissant's poetics, which is based on his analysis of Caribbean creolization and of globalization, lies not so much in its emphasis on the world's interrelatedness and connectedness itself, as that notion has been developed in many environmental discourses, though often in problematic ways. The significance of Glissant's poetics is rather located in its complex, critical approach of the notion. The following chapters will offer an interdisciplinary discussion of some of the problematic aspects surrounding the concept of connectedness, but I can here already touch upon the unease with the concept in philosophy of science. For example, in his evaluation of the philosophical debates around environmentalist discourses from the 1970s onward (most explicitly Arne Naess's "Deep Ecology") that assume a "fundamental connectedness/relatedness in the natural world that encompasses Homo Sapiens" (Kirkman, 2002, 4), science philosopher Robert Kirkman notes that a model of unified totality is destined to remain out of reach (152), in spite of the hopes of many environmentalists. This conclusion is nothing new for the Caribbean theorists of creolization and interrelatedness, in which ambiguity and multiplicity have been keywords. Glissant theorizes the world as unknowable, because it is chaotic, unpredictable, and contradictory. This does not mean that Glissant gives up on the attempt to grasp this totality. Quite the contrary. In an exploratory essay on different efforts to create a new, nontotalizing, nonrepressive, nonexclusive discourse on the universal, Natalie Melas explains that such discourses are as much-needed as ever, for "the notion of the degeneralized universal" can be seen "as one aspect of contemporary globalization's cultural potential" (2001, 135). Giving up the desire to speak about the world-totality would amount to giving up on working toward a nonconservative, nonimperialist theory of the transnational world. The question remains *how* to address that totality, as Glissant's poetics of the world-totality is by no means the last word.[19]

Glissant's poetics is, nevertheless, relevant to my discussion about the role of literature in environmentalism, as it argues that the interrelated world can never be completely known. Glissant's insistence on the opacity of space captures that insight very well. This is also one of the main insights in Mootoo's novel: the world can never be known by watching, naming, and classifying it. The words of Otoh's mother (quoted earlier) underline that insight, as do the images of insects on the novel's pages, as does Tyler's hesitant style of narrating the story. In other words, the privileged way of knowing is neither verbal nor visual.[20] But that does not mean that one cannot interact with the world. If Glissant calls for the incessant *imagination* of the world, then Mootoo, in line with Merchant, calls for an ongoing, subtle *interaction* with the world—an interaction subtle as the insect touch, open to the other's response.

Cereus is close to ecological feminism in the subtlety with which it unravels the intertwining ideas about nature and social issues (gender inequality, sexuality, race, colonialism). But the question that seems to drive the novel is not how it might imagine the world truthfully, but rather, which poetics is most helpful in *relating* to the often painful, intrusive, globalized world—to find ways to *live* in it. Through the narrative of an abused daughter who tries not to intervene in nature's business, Mootoo's novel depicts a pathological attitude to nature, an attitude dictated by trauma, resulting in the inability to interact with one's surroundings and in the creation of the ambiguous yard. The work as a whole insists on the imperative to interact with one's environment, not in a "natural" positive response to the world's connectedness (for that connectedness can as easily be destructive as it can be healing), but in an extremely careful, empathic way, opening oneself to the unpredictable sensuality of the touch that does not obey the patriarchal, erotophobic (Gaard, 1997,118) scripts of colonialism.

Cereus, then, takes as its point of departure the assumption that interaction is the key to a productive lifestyle in a damaged postcolonial world. This interaction inevitably implies border-crossing, and will lead to syncretism. The novel shares many of the assumptions of ecological feminism, even if it is critical of the ecofeminist belief in a harmonious, reciprocal relation with nature—it is much closer to the queer ecofeminism advocated by Gaard and Mortimer-Sandilands. Instead of endorsing the tenets of system ecology, in which the natural environment is seen as ultimately autonomous and best left alone to develop its own stable ecological system, *Cereus* shows more affinity with an evolutionary ecological approach, where the environment is seen as always unstable, always changing, and responsive to social and environmental events (Keulartz, 1998)—without embracing the focus on procreation and heterosexuality implicit in many evolutionary approaches. *Cereus* offers neither the plot Merchant desires nor a vision of a definitive return to an undamaged, original environment. It does offer the inspiration to think through the complicated global and local contexts of environmental sustainability. More strikingly, perhaps, than its capacity to critique colonial and national narratives of nature, is the powerful way in which it helps to uncover the very real and deep-seated anxieties and interests that shape society's attitudes to border-crossing, especially in relation to sexuality. The most valuable aspect of this novel is its ability to let us *feel*, that, ultimately, nature always exceeds all social discourse that tries to contain it.

NOTES

1. I understand *alterity* here as "outside or prior to conceptualization and understanding" (Malpas and Wake, 2006, 144). Alterity is not synonymous to the Lacanian other or Other, nor to the "other" as the alternative pole in the self/other binary. Malpas and Wake add that in phenomenology, "alterity can never be situated within a horizon of familiarity" (144).

2. Let me add that Susan Stryker and Stephen Whittle (2006) differentiate queer studies and transgender studies; the former would focus on sexual object-choice, and with that on desire and sexuality, whereas transgender studies would adopt a broader view of nonheteronormative positions, and be more concerned with embodiment and identity (7). It is the concern with embodiment that makes transgender studies so relevant for ecocriticism.

3. See Kutzinski for a sharp analysis of the psychological reasons behind this manipulation of time, and the role of repetition in Mala's way of dealing with her trauma.

4. In spite of the critique implied in this statement, Blair shows that women writers can make very good use of the metaphor too, by appropriating it and using it ironically.

5. Seymour describes Tyler as transgender, instead of transvestite, as she prefers the "current community-derived umbrella term of 'transgenderism'" (30). In this, she is following Susan Stryker's broad definition of transgender phenomena as calling into question "both the stability of the material referent 'sex' and the relationship of that unstable category to the linguistic, social, and psychical categories of 'gender'" (2006, 9). Indeed, toward the end of the novel, Tyler's cross-dressing shapes their own experience of their own body, so that the relative stability of what was presented as a male body is completely undone. As the cross-dressing is an important narrative thread throughout the novel, I will nevertheless describe Tyler in relation to that practice. In addition, I will follow Leslie Bow's example by allowing myself to be guided by the name Tyler offers to the readers at the start of the novel, a name that signals his masculinity. It is only at the very end that Tyler, while reporting an intimate conversation with Otoh (then Otty), reveals the significant detail of now being called Ty.

6. This is symbolized in the almshouse yard, the well-ordered garden of the Thoroughlys, and the Royal Botanical Gardens. See also Casteel.

7. De Certeau (1984) differentiates tactics from strategy. A very concise summary might be that while a strategy is developed from a position of power and control over a territory, a tactic is deployed by those who are not in power, and are navigating the space of the other (36–37).

8. Jamaica Kincaid, for example, in *A Small Place*, (1988) suggests that the excessive beauty of the Antiguan landscape is related to its exclusion from world history. Surinamese-Dutch writer Astrid Roemer shows that corruption is part of Caribbean nature. If women writers are unwilling to accept utopian imaginations of Caribbean landscape, it is often because they know all too well where these images come from.

9. The last statement is borrowed from Burke and Gallop, 1985, 112.

10. Although hybridization and creolization are both concepts referring to processes of mixing and merging, they cannot be considered as synonyms. See, for example, Purdom (1995) and Hoving (2001b) for an evaluation of the concept of hybridity, and see Shepherd and Richards (2002) for a discussion of creolization and creole.

11. Proust's work offers a good example, highlighted in an essay by Samuel Beckett, who states that in Proust "[h]omosexuality is never called a vice: it is as devoid of moral implications as the mode of fecundation of the *Primula veris* or the *Lythrum salicora*" (Beckett and Duthuit, 1965, 89–90). I would like to thank Dirk van Hulle for presenting me with this example. Bauer (1997) offers a more elaborate analysis of plant metaphors in male French literary writing about homosexuality. A comparable use can be found in Caribbean gay writing, such as Lawrence Scott's *Aelred's Sin* (1998); see chapter 4 of this volume.

12. Flowers are apt images to explore issues of sexuality. The work of other Caribbean writers shows that images of flowers, plants, but especially trees, can be used to question colonial assumptions, such as those about identity as autonomous. See Tinsley (2005) for a great analysis of the 'language of flowers' within the context of Surinamese mati-work. See Licops (2001) and Mitsch (1997) for an analysis of tree metaphors in Caribbean writing. See also the next chapter.

13. As Casteel (2003) has argued, the novel also relates how the cereus was transplanted from one garden to many other gardens.

14. This remark does not imply that the novel shares a Deleuzean aesthetic. Its love of horizontality, thematized in the dwelling on surfaces and indexicality, is balanced by its insistence on secrets that must be acknowledged before life becomes possible.

15. As border-crossing is here imagined as relating to race, generation, and so forth, as much as to gender and sexuality, I offer a different reading than Seymour (2013), who primarily reads the novel as helping to effectuate the 1990s shift between "the older sexological model of 'transsexuality' to the current community-derived umbrella term of 'transgenderism'" (30). Though I am fully convinced by Seymour's point, I argue that the novel's project is even more ambitious and chooses to adopt a more encompassing intersectional frame.

16. Compare to Leslie Bow's (2009) analysis of a transgender person's statement in Marlo Poras's documentary film *Mai's America* (2002): "I don't feel that I'm right here. [. . .] I feel like I'm out of place" (94). Bow defends the benefits of "drawing an analogy between transracialism and transgenderism" (100), while being aware of the drawbacks.

17. A big thank you to Elizabeth DeLoughrey for suggesting this conclusion to me.

18. Here, I offer a different emphasis from Casteel (2003), who reads the garden more negatively, as a "Gothic landscape" of rot and decay, "not only a vegetable graveyard but also a human graveyard" (19)—as the macabre unmasking of the desired "timeless garden" as a myth (20). I prefer to read in this garden an alternative, queer imagination of nature too—one that recognizes the inevitable intertwining of sensual and erotic delight on the one hand, and the abject on the other.

19. As I will discuss in detail later, Peter Hallward (2001) suggests that Glissant's later work increasingly represents a so-called unproductive "singular" approach to the world, in which he (like Deleuze) assumes that reality is the expression of one self-actualizing principle. As this principle realizes itself independently from any ethical or political context, it is hard to see how such a singular approach—in Hallward's slightly unusual sense—would help in analyzing specific political conditions. See especially chapters 4, 5, 6, and 10.

20. Kutzinski discusses the novel's representation of spaces in which "existing does not mean being visible to someone else, as it does in Hegelian paradigms of selfhood" (Kutzinski, 2001, 179).

Chapter Three

The Erotics of Mimesis

"Art makes order from disordered shards and pieces," says Timothy Reiss (2002, 358), with reference to the fragmented Caribbean. Prior to Reiss, Derek Walcott, too, had referred to this function of art, as have many critics in this field—from the discontinuity and fragmentation of the Caribbean springs the "need for Caribbean artists to give themselves to the 'restoration of our shattered histories, our shards of vocabulary, our archipelago becoming a synonym for pieces broken off from the original continent'" (Walcott 10–11, qtd. in Reiss, 355). Botanical metaphors appear to play an eminent role in this project of unification. After a consideration of such botanical images in, among others, Jamaica Kincaid, Edward Kamau Brathwaite, Heberto Padilla, Paule Marshall, and Alejo Carpentier, Reiss recapitulates:

> Flowers are one mark of a new making, of joining, of setting new histories in their chosen place. Ultimately, for Glissant as well, they define "the patience of landscape . . . not saturated with a single History but effervescent with intermingled histories, spread around, rushing to fuse without destroying or reducing each other." (1989, 154, qtd. in Reiss, 353)

One could write a history of the many metaphors that, like these flowers, are used to imagine the irretrievable fragmentation of the New World. Some are musical (Brathwaite's jazz), others culinary (aijaco). The melting pot and the mosaic have gained currency. Mestizaje, créolization, métissage carry their own regional histories. As is to be expected, these metaphors all offer different theories of the quality, degree, and manner of the unification of the diversity they address. Some suggest complete assimilation, where others emphasize the syncretic nature of the coming together of incommensurable aspects which can never be translated into each other to form a new organic whole. Many of them conceptualize the irreparable syncretism, artifice, and instability of the region rather than the effort to heal this fragmentation. Thus, they suggest that the way in which Caribbean art creates order should not be simply understood as a movement *from* fragmentation *toward* wholeness. Instead, it should be theorized as an exploration of the *many* ways to relate to fragmentation.

Building on this history of coining metaphors of fragmentation and unification, Reiss considers the European imagery of flowers, going back to "the younger Seneca's familiar analysis of mimesis as imagined in bees gathering juices and flavors from flowers to turn them into a new 'savor, in such a way that even as it shows whence it came, yet it appears as other than whence it came.'" This becomes a metaphor for a new consciousness which "impress[es] a new, its own, 'form on all the things which it has drawn from what we may call its examplar, so that they/ join in a unity' (*Ep*, lxxxiv:2.276–85)" (352).[1] In this sense, bees and flowers are implied in a developing European discourse of wholeness out of diversity, which was subsequently appropriated and subverted by Caribbean writers. Flowers could then become, as in Césaire's poetry, "a violence *against* European poetic forms, against European History" (352). But they also represent an older indigenous violence, as Wilson Harris claims. Reiss evokes how Harris imagines old "'Tetoctitlan . . . upon which flowers grew inscribed with dazzling blood'—out of that violence, the flowers forge a fresh experience of place, one that is anti-European to be sure [. . .] but which is more importantly and simply *not* European," while it still relates to Europe (352).[2, 3]

The unifying force of flowers is the topic of this chapter. A flower was central to the novel discussed in the preceding chapter. Flowers are also central to *Aelred's Sin* (1998), the second novel by Trinidadian writer Lawrence Scott, which was awarded the Commonwealth Writers' Prize Best Book in Canada and Caribbean (1999). In both novels, flowers function as means of (re)connection. In *Cereus Blooms at Night*, such healing forms of reconnection and border-crossing become possible through (among other things) a recognition of sameness in others. One of the most effective critical devices in the novel, the comparison of people to plants and animals, allows for a liberating, queer understanding of sexuality and race. Thus, a hermaphrodite, insect-pollinated flower was made to bridge the colonial divisions between "races," genders, and sexualities. However, *Cereus* is not only dependent on the notion of *mirroring* (as if people and plants mirror each other, as if people would or should copy plant behavior—as if plants would be the hidden reality beneath people's behavior), but also, and even more so, on that of *connecting*. The notion of the careful touch is explored in much more detail than that of similarity. I differentiate the two strategies, connecting and mirroring, with some emphasis, because they represent two aspects of the double-sided concept of mimesis as analyzed by anthropologist Michael Taussig (1993). According to Taussig's view, mimesis refers not just to visuality (copying, mirroring—which is the traditional understanding), but also to corporeality (contact). If *Cereus Blooms at Night* focuses on the corporeal aspects of mimesis, *Aelred's Sin* ponders the potential of *visual* mimesis as a means to reconnect the (post)colonial world. For reconnection is also this novel's project. As the main protagonist (Aelred/J. M./Jean Marc) puts it: "There are these disjunctions in my life which I must yoke together if I am to survive" (433).

However, the traditional, visual strategies of mimesis (identification, mirroring, imitation, copying, erasure of otherness) have been subject to severe critique. As Hansen (1999) suggests in an essay on Benjamin's notion of innervation, imitation (in contrast to play) has been associated with fascism (that is, in the form of semblance, illusion), and thus with manipulative authority and universal, perpetual truth (321–29, esp. 324). In chapter 4, I will show that strategies of identification and mirroring are radically dismissed in Kincaid's work. In this chapter, I will discuss the inherent (and perhaps intended and welcomed) paradoxes in Scott's use of the visual strategies of mimesis for the project of postcolonial unification. The point here is that mimesis entails a loss of self. I want to suggest that Scott puts the many images of flowers and plants at work to find a solution for this problem, one that cannot be solved at the level of the plot. It can be solved, however, by calling on the strong spiritual and erotic traditions that flowers carry with them, and that offer an alternative discursive frame through which to imagine postcolonial connectedness.

DO LOOKS KILL?

Aelred's Sin is structured as a frame story containing several other, intertwined narratives. The first narrative is set in the 1980s, when a Caribbean man from a French creole family reads his deceased brother's journal, to learn about his life and untimely death. His dead brother, Jean Marc (J. M.), is the main protagonist of the second narrative, which begins in the 1960s. Fascinated by the life history of the twelfth-century abbot Aelred of Rievaulx, whose life becomes a continuous point of reference in the novel, J. M. moved to England to live as a monk. A third and last narrative thread consists of references to an unnamed black boy on a painting in the monastery, for whom Jean Marc invents a life, inspired by childhood stories about an escaped African boy and a stone in the monastery's graveyard.

Let me begin by discussing the intense forms of identification, through which these characters connect. This is the aspect of the novel that draws immediate attention and has formed the rewarding object for most critical analyses of the novel (e.g., Donnell, 2006, 212–13; Rahim, 2005, 35; Wroe, 2004, 431–34). The key scene occurs when Jean Marc, now Brother Aelred, is dusting a painted portrait of an early owner of the estate Ashton Park, which has been turned into a monastery. When Aelred discovers a small black boy kneeling at the duke's feet, his thoughts go back to his own youth in the Caribbean. "Then he saw his own face reflected in the glass of the portrait. His face was superimposed upon that of the boy whose face shone from beneath, so that the black face seemed to be his own" (78–79). While gazing at the boy, he remembers the reproaches of his young black childhood friends, that he, as a French Creole, tended to act superior. Then he remembers the story his old nurse told him about the runaway slave boy Mungo. The next moment he sees his best friend among the monks, Benedict, leave through a door. It slowly dawns on him that this friend had been staring at his legs, which he had accidentally bared during his chore. However, the eroticizing of the scene was already foreshadowed, when Aelred was struck by the plaintive admiration in the painted black boy's eyes, which doubled that of the duke's dog at the other side of the painting (78). Amidst the confusing reflections of the light, Aelred finds that the boy is wearing an iron dog collar. He realizes this at the very moment when he recalls his friends accusing him of acting racially superior, as a light-skinned Creole—a memory that is now overlaid by his identification with the slave boy. The impression the small black boy makes on him is so powerful that he is confused about his own observation of his friend Benedict's presence. In the midst of this confusion he remembers that he was told that the runaway slave boy Mungo had a scar on his neck. A few pages later he returns to this discovery, which, as though it were a primal scene, is now eroticized explicitly through the words of his old nurse: "Mungo have a scar right there on his neck. Like when you sweetheart give you a kiss on the neck and you could see the bite. But is no kiss" (99). The scene before the painting receives its full intensity when he remembers (or rather, "sees," as the text has it) a heterosexual childhood friend who, years ago, had also been watching his legs. His memories are still confused; he has trouble to forget both the face of the black boy on the painting, as well as the face of his fellow monk Benedict (82). Immediately after evoking this confusion, the text informs its readers that Aelred's (explicitly) heterosexual childhood friend had told him that he wanted to kiss him as if he "were a girl" (82).

As Donnell (2006) points out, Aelred's "highly transgressive identification [. . .] crosses an almost originary divide in Caribbean societies between those whose ancestors were slaves and those who were slave-owners" (213). She places this identification next to Aelred's identification with the abbot Aelred of Rievaulx, who pleaded for the recognition of spiritual love between monks, and to Aelred's brother's efforts to identify with his homosexual brother. Donnell understands these "recognitions of self and other," or "instances of empathic mobility" as a hopeful indication of the possibility to transgress "the boundaries policed by both national and sexual identity politics" (213). Emily Wroe (2004), more radical yet, reads the portrait as "a gateway through which synchronous dialogues at a primal level emerge" (433), differentiating, in the wake of Wilson Harris, between a historical and a primal level. There, Aelred is able to enter what Harris calls the "womb of space"—his imagination of a universal consciousness, through which he can connect to and identify with otherness (see also Emery, 2007).

Poet and literary critic Jennifer Rahim (2005) is less optimistic, however. Like Donnell and Wroe, she acknowledges that Aelred "comes to recognize Jordan (the painted slave boy, IH) as his sexual and racial Other, whose history reflects his own experiences of sexual violation by his homophobic schoolmates and the racialized construction of his body by his fellow English monks" (35). Rahim, however, is more attentive to the sadism (and, I would add, masochism) at the heart of these juxtaposed narratives. "Scripts of sadistic sexual abuse by colonial masters, consensual same-sex love and homophobic violence therefore crisscross, corrupt and interrupt each other as Lawrence (Scott, IH) attempts to undo narratives of homosexual negation in order to create a redemptive construct of Jean Marc's life" (36)—a life that, as she points out, is nevertheless made to end in a tragic early death due to AIDS. It seems to me that Scott's narrative is in the first place *exploring* the dark, destructive sides of sexual and racial politics, while nevertheless trying to relate to them sensually, instead of merely seeking redemption—though the latter dimension is certainly present. But the narrative is at least as much concerned with a (feared and desired) loss of self as with the creation of a self.

In her probing and detailed study of the role of visuality in Caribbean literature, Mary Lou Emery (2007) analyzes "the significance of vision to one of the major intellectual shifts of the twentieth century—the refiguring of identities across national, racial, and cultural boundaries" (1). She frames her analysis with the help of a close-reading of David Dabydeen's both fictional and non-fictional work on the prints and paintings by eighteenth-century painter William Hogarth. At the heart of Dabydeen's work (including his 1999 novel, published a year after *Aelred's Sin*), we find an image that doubles that found in Scott's novel: an eighteenth-century art work that represents a small black boy in the presence of his colonial masters.[4] David Dabydeen shows how Hogarth's art brings out what implicitly marks the usual representations of black people in that period: the paradoxical alignment of commerce and art, of slavery and aesthetics. Referring to Simon Gikandi's work on the same subject, Emery emphasizes the "centrality of this conjunction (slavery and aesthetics, IH) in the constitution of modernity" (7). She traces a tradition that begins with Hume's comparison between artists of African descent who merely imitate, and the European artist who is truly creative, a "modern subject-who-sees" (8). That tradition passes through the work of Kant and Schiller, who developed the link between a modern, rational subjectivity, aesthetics, and freedom: "In the practice of aesthetic judgment, defined as both disinterested and universal, the individual constituted himself as free personally and also socially within an emergent public sphere. [. . .] Through the eye, the subject constitutes himself as free" (9). This specific way of looking is described as the "'higher sense' of distanced, contemplative sight" (9). In the twentieth century, David Spurr (1993) will point out how that distanced sight is connected to appropriation: "This 'arm's length' view takes possession of social reality itself and can become a 'mode of representation by which a powerful culture *takes* possession of a less powerful one'" (59). For him, in this context, "aestheticization *is* colonization" (18). David Lloyd underwrites this statement by insisting, in Emery's words, "that the visual structure of racism is not based on an antagonistic recognition of visual difference, but on establishment of a subject-who-sees as the universal human Subject" (Emery, 2007, 10).

This free modern subject advertised his privileged position by flaunting the art he had bought, a possession that, while being bought with money made in colonial trade and slave lablor, "appears free of such economic and moral constraints" (11). At the same time, the picture itself will point at its origins in other people's lack of freedom, by including black people within its frame. As Gikandi has it, "the categories of the aesthetic and slavery 'operated within the same economy of discourse'" (qtd. in Emery, 10). Commerce and aesthetic taste are the two means to become part of society. Gikandi discusses this predicament in relation to a white Creole, whose family had become rich because of their plantations. This man, who is seen as a relative outsider to English society, tries to solve the paradox by turning into an aesthete. Lawrence Scott's light-skinned Creole hero, Aelred, chooses a comparable path. Instead of becoming an unambiguous aesthete, however, he chooses to enter the world of religion, understanding that world first and foremost as an aesthetic sphere—albeit an aesthetic that is not based on the distanced, disembodied eye described by Lloyd, but on an intimate, reciprocal, synaesthesia. In this, he finds a kindred soul in Derek Walcott. As Emery points out, Walcott too finds a spiritual experience in art (197–98). Moreover, like Scott, he also relates it to the erotic.[5] This aspect, neglected by Emery's analysis, is paramount to Aelred's narrative.

For in the gaze's production of subjectivity, the subject is also *gendered*. Freedom is not just a white, but also a heterosexual, masculine prerogative. The objectification produced by the imperial gaze is also gendering, sexualizing, eroticizing. Like other critics, I argue that Aelred's contemplation of the eighteenth-century painting in the monastery can be seen as deconstructing the imperial, objectifying gaze. But it remains to be seen what exactly makes this deconstruction possible. Is it interracial identification? Is it merely the identification with another victim of sexual abuse? Is that form of connecting in itself already adequate to undo colonial divisions? Bearing in mind that any imagination of connectedness and copying also implies a certain imagination of sexuality, I propose to answer that question by analyzing the workings of sexuality in this scene.

We have seen how the mirroring scene evokes racial power differences that are consequently subverted by sexual difference. Within the intensely racialized/sexualized frame established by this central scene, the positions of viewer and viewed, pursuer and pursued, subject and object, are continuously being exchanged. As the monk who stared at Aelred's naked legs is described as "lost in his staring" (81), whereas Aelred is pictured as confused and uncertain about what he sees, we can argue that the eroticizing gaze is never fully possessed by someone. As such, it cannot be equated with the objectifying colonial gaze. In addition, it constantly shifts between the characters. Objectivity and subjectivity shift to such an extent that even aggression may turn into sexual submissiveness, or that perpetrators may be unmasked as led by sexual desire. A young English monk nags Aelred about his dark skin (161, 167). The reader will not be surprised to see the racist pest become Aelred's eager lover, since this script also underlies the homophobia the protagonist encountered in the Caribbean: hatred covers up desire. The most violent homophobes were possibly the most jealous of same-sex love, and the supreme instance of homophobic violence consisted of a homosexual gang rape. Sexuality and race are intertwined to the point where racism is almost equated with (sadistic, violent) sexual desire, a point already amply explored by critics ranging from Frantz Fanon to Ann Laura Stoler and Cynthia McClintock. Tellingly, in this novel, iron chains leave marks like kisses, and the red rose Aelred hands to his future lover wounds his finger (166).

Donnell makes an illuminating point when she sees in the mirroring scene a "bizarrely inverted representation of Fanon's *Black Skin, White Masks* (1952) which similarly marks the desire to be in two places, even two identities at once, that Fanon analyses as a condition of the colonial subject" (213). Fanon concentrates on the black desire of whiteness, whereas Scott's scene stages a light-skinned Creole desire for/identification with blackness, and a white desire (in the staring fellow-monk) for the Creole body (Aelred). The inversion is not so bizarre, however, when we relate it to Homi Bhabha's analysis of colonial mimicry and the colonial desire to *be* mirrored by the colonized, a desire which is closely related to the colonial or European desire to *become* the racial other, that lies at the heart of so many instances of orientalist and colonial imaginations (Bhabha, Boone 57, Fanon, JanMohamed, Mercer, Thomas).

The inevitably ambiguous colonial representations of otherness form a rewarding subject for artists and writers, as the problem of imitation is already exposed and questioned in colonial art itself. If a monkey or little black servant copies the master's dress, the master is already mocked, or is unmasked as imitating others, too. His authority is questioned by the mere fact that he is imitated—as he asked to be—and his authority is increasingly endangered when his identity is appropriated by ever more others. By way of example, Emery refers to a novel on a painting by Wilson Harris, who plays with that principle quite effectively. The master (Paul Gauguin) is doubled again and again in other characters named Paul, who lack his authority (206). For this reason, master Gauguin loses his absolute authority as a model. The same slipping and doubling works in *Aelred's Sin,* a novel that testifies to a comparable pleasure in linking and connecting. As I have already argued above, in this novel no one seems to really possess the masterly colonial gaze. Visuality is not authoritative but always destabilized, as it is embodied in a desiring viewer who may lose himself in the desiring gaze.

However, this well-known argument on colonial mimicry does not solve the question of why the gaze in the scene before the painting in *Aelred's Sin* would effectively *subvert* the colonial relations. The colonial mirroring game in itself does not necessarily result in the healing or deconstruction of the racial/sexual divides. As it merely repeats the colonial dynamics that installed the divides, the only thing it would be able to do is unveil, criticize. Most critics, however, agree that this scene has a more powerful subversive effect. Remarkably, it is through the explicit eroticization of these divides (that is, the intensification of an erotics that was already implicit) that the destructive effects of these dynamics are undone. The gaze that organizes the scene is not the imperial, objectifying gaze of the master. Nor is it the plaintive, admiring look of the slave boy and the dog on the painting from below. It is, rather, the doubling, ambivalent, homoerotic gaze. The homoerotic gaze, as Kobena Mercer (2003) explains, contains "an element of erotic objectification, but it is reversible. [. . .] The gendered hierarchy of seeing/being seen is not so rigidly coded in homoerotic representations" (194). In this context, one may be objectified, or objectify others, but never permanently. The homoerotic gaze that objectifies also invites response.

It is this reversible homoerotic gaze that subverts. One could well wish to problematize this gaze for the very reason that it follows the colonial logic of desiring/fearing the racial other. Nevertheless, it seems to me that the novel succeeds in negotiating these risks. To make myself very clear: I do not argue that Aelred is eroticizing the black boy on the painting (though he may be doing so). What I want to argue is that the scene in front of the painting (and *in* the mirror that the painting becomes) follows the dynamic of the reversible homoerotic gaze in a way that challenges the kind of rigid objectification at the heart of colonialism. Within this space, Aelred is able to invite the black boy to respond—and he succeeds in doing so by giving the boy a story. This effort to give the other a voice counters the colonial logic. In addition, this gaze is not merely interested in interracial identification. The mirroring does not only concern racially different characters. *All* (male) characters may merge. The name of Aelred's first lover, Ted, is echoed in white Edward, Aelred recognizes his own face in that of Mungo/Jordan, but also in that of white Benedict, until, after a dramatic moment in the monastery, he finally sees the faces of all his young and old male friends merge (409). It is therefore not necessarily a shared history of victimization that would decide whether a face can be mirrored or not, as Wroe suggests. The fact that, for example, Benedict is also part of the chain (when he is still secure in his position as senior monk, before he falls victim to self-willed starvation), points at the inclusive force of this strategy.[6]

By mobilizing the gaze as a force that produces racial and gendered subjectivities and objectivities—that is, by mobilizing psychoanalysis—Scott's writing shows that a critical negotiation of the gaze is not necessarily geared toward the creation of an emancipated Caribbean visual subjectivity, it also explores the (desired) loss of the self that is implicit in the (also objectifying) gaze. This possible loss of self is an inevitable dimension of the open, reversible gaze, in which one risks to be gazed at in turn. This willingness to take on the position of the object may seem harmful, but it nevertheless has tremendous political force. Colonialist and nationalist discourses are violently repressive because of their anxious efforts at stabilizing unstable, threatened subjectivities, at the expense of (black) women, for example. These efforts are structured by a masculinist, heterosexual logic that abhors, in turn, all efforts at *being* objectified. By patiently exploring the many violent *and* sensual aspects of the lived experience of objectification, Scott's novel is addressing that very anxiety.

It is true, the risk remains that this work with analogies can lead to the repression of difference. In explorations as these, in the risky, shady area of interracial desire and anxieties, there can never be a fixed boundary between an "imitative masquerade" and a successful "ethical and political alliance," as Mercer admits with great insight (197). *Aelred's Sin* seems to explore that unease too, by articulating the drawbacks of incessant mirroring.

One of these drawbacks can be explained with the help of Lacanian psychoanalysis. One could read the tragic ending of Aelred's narrative as the suggestion that he is stuck in the colonial Imaginary of an eternal, sadistic/masochistic game of mirroring that leads to a very definite, irreversible loss of self in death. Aelred suggests in a letter to a friend that it has always been his desire to have a *brother* (434), which might situate his nonoedipal sexual desire within the Lacanian Imaginary. This space would be marked by the dialectic of love and hatred for the other, who functions as one's dangerous mirror image. Even if the text is fascinated by mirrors, it also seems utterly aware of the dangers of identification. The fear of being devoured by one's own or another's passion is very present. "He was frightened because he [. . .] knew that his feelings [. . .] could engulf him" (22). Later, when fellow-monk Benedict declares his love, Aelred realizes: "He felt he could not cope with the knowledge [. . .] it was so powerful and good that he immediately wanted to reject it. [. . .] He wanted to hurt Benedict for the first time. [. . .] Suddenly he felt powerful. It felt cruel. He wanted to hurt Benedict" (141). However, the novel also acknowledges the erotics at the heart of such a loss of self. The mirroring game is intensely erotic, and through the style of writing itself, the reader is drawn into that enjoyable, desperate pain. The many repetitions, the interweaving of endless comparisons (as in the mirroring scene, although the novel counts many, many more examples) to postpone conclusions, the questioning of visual observations—all of these stylistic devices help to intensify the dizzying sense of being caught in the boundless, erotic, anxious, always frustrated quest for identification that is also the quest to lose oneself, until it harshly ends—for Aelred and his lovers—in death. This desire can be linked to the *mystic* need to lose oneself to a transcendent authority, something greater than the self—which is the very definition of the masochist desire. Aelred himself calls this an "*erotic mysticism*" (313). Aelred's sad brother, the first reader of Aelred's narrative, discovers in these unhappy lives a "strange synchronicity . . . a macabre pattern . . . denial, punishment, death. The pattern becomes compelling" (307), a pattern that the novel itself, in its writing style, repeats. Robert Scholes might comment that this narrative is very good indeed at the sadistic game that every story plays with its readers (Scholes, qtd. in De Lauretis, 1984, 108). Indeed, here we find another indication that the novel is not merely concerned with the emancipatory wish to create a stable, social self, but also with what Leo Bersani (1987) would call the shattering of self in sex.

I discern two different levels in this text. On the one hand, we have the embedded, dizzying narrative of Aelred's quest for (spiritual and emotional/sexual) surrender, which ends in death. On the other hand, we have the identitarian framing narrative of Aelred's brother Robert's quest to understand his brother, which leads to a more traditional emancipatory story of the quest for a representable, social self, including a social critique of the society that denied Aelred a self. For it is society itself, including the church, that refused Aelred a father. None of the father figures—abbots, older monks—in the novel knows how to sensitively, affirmatively respond to this need for a loving father. The exception is the twelfth-century abbot Aelred of Rievaulx, and one might argue that this thus doomed Aelred to remain in the closed circle of the Imaginary. At the end of his life, Aelred complains how his love for a man had "no community to give it recognition, no traditions, no laws, no rules. They allowed us nothing but secrecy and anonymity" (435). In other words, they were refused a symbolic order.[7]

Scott's novel is well aware of the risks of this play with analogy, mirroring, and connecting. It seems to try to balance the identitarian, emancipatory narrative that offers a stable sense of self and a traditional notion of interconnecting, with a narrative in which that sense of self is shattered. The novel has been criticized for not offering a satisfactory narrative solution to these tensions. However, the tensions explored by Scott are indeed irresolvable, as sexuality itself is marked by tensions between the desire to lose the self and to assess the self. Nevertheless, though the plot is not able to solve these paradoxes, they are taken up and brought together on another level: that of the text, and the choice of metaphors.

With the help of these metaphors, all of which refer to nature, the novel is able to offer a frame for the imagination of the self that exceeds the identitarian or psychoanalytical. Indeed, the mirroring-game between self and other can only be resolved through recourse to a third term. As Scott has deconstructed the Law-of-the-Father and his All-Seeing Eye, he must adopt a completely different symbolic discourse to replace it. Instead of the Paternal Eye, he addresses sensual nature as a third term. The patriarchal transcendence of the signifier is unattainable, so the nontranscendent, sensual ideal of fraternity takes its place. When that cannot be obtained, the nontranscendent, embodied reciprocity that can be found in some notions of nature takes its place. Here Scott turns to the rich symbolism of flowers that might help him to stabilize and deepen his own aesthetics of linking and mirroring, since flowers have, as Reiss pointed out, for centuries signified the force of connecting, merging, and interlinking. The next section offers an analysis of the way in which Scott uses nature, and especially the image of flowers, to supplement his anxious, potentially endless game of mirroring.

INSTEAD OF THE FATHER: FLOWERS AND FLESH

Scott's landscapes are overwhelming. Emily Wroe reads them, inspiringly, through Wilson Harris's concept of the womb of space. Scott's notion of synchronicity evokes a comparable image of a universal unconscious where the temporal order is dissolved, and which gives access to all cross-cultural imaginations. Wroe sees nature as offering Aelred an opportunity to escape from the oppression of social life. Aelred even recognizes in his imagined, marooning alter-ego Jordan the desire to turn into an animal or plant to gain his freedom (437).[8] Wroe is right in marking this moment as a turning point in the narrative, as it gives Aelred, through an identification with Jordan, who chose to escape objectification by running away, the conviction to claim an identity as a man who desires another man. But this short excursion in the realm of nature that might offer an alternative to the colonial world soon passes. The subtext of nature that forms a constant background to the novel is not the sublime landscape,[9] but most often appears in the continuous references to flowers. It is to their significance that I will turn now.

Images of flowers function as complex knots, where the different discursive strands in the novel are tied together. First, they refer to the theme of same-sex love, inserting the novel in "the no longer closeted gay floral tradition" (Bauer, 1997, 81). The abundant references to flowers in the narrative resemble the way in which flowers in *Cereus Blooms at Night* were used to reconsider the notion of natural sexuality. Scott's novel's constant evocation of a variety of scented flowers ties in more explicitly with a *male* literary tradition, however, in which Baudelaire's *Fleurs du Mal* looms large.

> The gay body['s] fictions, depictions, and interdictions curiously fleshed out in a dialogue with natural history become natural science stem more from botany than zoology. As gay figures begin to emerge in nineteenth-century French literature, they are nosegays whose bouquets are perfumed and dandy floral offerings that flirt with a certain language of criminality and real and artificial flowers. (Bauer, 1997, 65)

The suggestion of an intimate relation between language and flowers in the quote is important, as I will demonstrate below. So is the slip from criminality to flowers—or, as Bauer sighs poetically, "the slip from a gay tissue of flesh to a libidinal swamp, to flowering evil" (65).

Anthropologist Jack Goody (1993) traces the central role of flowers in the French nineteenth-century literary tradition in general back to the immense popularity of a supposed secret language of flowers in Europe's East, made famous in the letters of Lady Mary Montagu in the early eighteenth century. Lady Montagu spoke of a Turkish habit to exchange secret messages with the help of objects (233). This became an Orientalist fascination, soon to be associated with transgressive sexuality, "a language known only to women, who 'invented it [. . .] and employ it as an amusement, or as a code for lesbian attachments'" (Hammer, qtd. in Goody, 234). As the market in exotic flowers from the East grew, books began to be published on a specific secret language of flowers. Such publications appeared in English translation after the first quarter of the nineteenth century (240). In the course of the nineteenth century, the genre was appropriated by religious writers, who endowed the flowers with moral meanings. However, it also found its way to those interested in occultism, and the genre was thoroughly subverted in the hands of Symbolist and other avant-garde writers (245–51). The genre's origins in Orientalist fantasies might be responsible for these literary flowers' capacity to evoke associations with secret, forbidden love. This explains why Scott can use them to evoke the same associations, referring both to their religious and their erotic connotations. At the same time, as Aimé Césaire mingled some of the elements of Symbolism with the dense significations he bestowed on the abundant botanical imaginary of his own passionate poetry, the complex imaginary of flowers as signifying sexuality, life, and corruption had already become an established part of Caribbean culture.[10] Through that

Caribbean heritage, Lawrence Scott evokes the full, ambivalent sensuality of an earlier, intercultural, exoticist tradition.[11]

One of the discourses addressed by Scott's flowers is religious. Goody (1993) argues that the early Christian church was critical of the use of flowers, as it frowned upon all earthly sensual enjoyment. But flowers were nevertheless slowly becoming part of monastic life, as the urge to admire God's divine creation led to the wish to decorate the altar (124). In addition, in the religious context, "flowers, like paintings, were used for instructional purposes." This was especially the case for the devotion of the Virgin Mary. The rose—symbolizing love, martyrdom, and death—was associated with her (155–56). However, flowers found their way into the church only when intercultural contact with the Eastern artistic love for flowers led to their growing popularity in Europe as well.

Scott's novel delves deep into this intercultural religious symbolic tradition and allows flowers to serve as intercultural mediators between England and the Caribbean, as well as lovers of different cultural backgrounds. Flowers are present throughout Jean Marc's story, as he moves into the English monastic life. He steadily takes "notes on landscape and flowers and birds" (61), comparing them with the Caribbean flora, allowing his brother to do the same after his death, when he reads Jean Marcs diary (45). It was Jean Marc/Aelred's task as the youngest novice to pick and arrange flowers for the shrine (162). He sees England as a garden (164, 166). In a dense scene full of premonitions, he hands his future lover a red rose, but it punctures the other's finger so that it bleeds (167). This gesture, then, is not only pointing forward to sexual bliss and the suffering that inevitably goes with it, but also to the essence of Benedictine monastery life. Saint Benedict himself, when he became a hermit and did penance, cultivated roses—"the flowers to delight the senses, the thorns to mortify the flesh" (Goody, 1993, 125). Thus, the erotic, gay floral tradition is interwoven with a religious discourse on plants and flowers that is already associated with transgression—and (masochistic) suffering.

Above, I referred to a gay floral tradition that develops in the work of the Symbolist writers of the nineteenth century and is continued by later writers. The Symbolist evocation of sensual experience was driven by the desire to create a transcendent perception of the absolute, a purpose that is also central to Aelred's life (though his desire at first follows the institutionalized path of the church). In a section about the ambivalence of the white European identification with blackness, Kobena Mercer (2002) suggests that the rebellious artistic Symbolist tradition still influences white artists who desire blackness as a means to escape modern society and high art:

> Mailer's white negro, who went in search of the systematic derangement of the senses, merely retraced an imaginary pathway in the culture of modernity

> previously traveled by Arthur Rimbaud and Eugène Delacroix in nineteenth-century Europe. There is a whole modernist position of "racial romanticism" that involves a fundamental ambivalence of identifications. (197)

If Scott's narrative calls upon the intercultural, transgressive symbolic tradition of flowers, it also evokes the following associations: a desire for interracial (white-black) identification, a sometimes self-destructive desire to lose oneself in a dizzyingly sensual world that is associated with racial otherness in a wild defiance of Western modernity, a quest for transcendence that dismisses mainstream religious discourses. Clearly, Scott is not interested in all aspects of this tradition, but he does mobilize some of them. Namely, the association between flowers, interraciality, transcendence, and the critique of institutionalized religion. Moreover, he is also interested in inviting a deeply sensual response to his texts, sometimes with the help of the device of synaesthesia. In this manner, the sensual begins to take precedence over the textual.

Scott's synaesthesia occurs most often by equating sounds (of words) with visual, tactile, or olfactory sensations relating to flowers, trees. Perhaps here we find Scott's most significant strategy, not the understanding of flowers as signs, but the opposite understanding of signs as flowers. The novel invites a perception of gaze, language, and identity that could be called phenomenological. This move occurs in close connection with the evocation of religious discourse, yet it opens another dimension within that discourse. "He found these high words, as he calls them, in the high hills, from the tall palms, from the palmistes. [. . .] He had begun to mix it all up" (129). Words come from trees. "Dom Maurus had white hands and beneath his pale skin the veins ran dark blue. Blood, like Quink ink" (18). And again: "The pelican's young pick at its breast. They drink blood. Images I'm reminded of. Drink this . . . words, words change everything. Blue veins the color of Quink ink. A poetry of pain" (115). This comparison, blood-ink, color-taste-touch (pain), is repeated again and again, including the reference to the brand name. There is no difference between plants, flowers, and trees on the one hand, bodies on the other, and language on yet another. A passionate game of comparing, identifying, and copying lets language and body, words and flesh, words and flowers collide and collapse.

The novel keeps evoking the sensuousness of scents and sounds, that always, Proust-like, will call up earlier sensual memories (see, for example, 48). English flowers and Caribbean flowers, English and Caribbean sensual and sexual experience, mingle and evoke each other. With the memory of the Caribbean scents comes the memory of Jean Marc's sexual relationship with his bosom friend Ted. All reminiscences of this sensual love are punctuated with memories of flowers and fruit, scents, taste—the perfume of the white lilies behind the latrine where they had sex—mingling with the smell of urine and shit (148), the juice of ripe cashew, the shade of the cocoa trees, the sounds of the cigale in the gru gru tree which drown their own cries when they make love during the siesta, amidst the sawing of the cicadas in the casuarinas (129).

It was the intense sensuality of cultivated nature that attracted Jean Marc to the monastery life in the first place, see pages 17–18. There is no doubt possible: "This was a paradise, a cloistered heaven" (18). Light and whiteness are central metaphors for this Caribbean space, in which a group of English monks searches for Divine Wisdom and perfection. The sensual delight in scents, tastes, and colors is evoked by both plants and by bodies. Jean Marc's childhood friends are pictured in terms of cinnamon, coral, tamarind seeds, incense, sandalwood—as if the text refuses the starkness of a binary imagination of colors, as in the dichotomy white/black (19). Even in this color symbolism we find a tendency toward nuances, overlap, linkages instead of stark, unbridgeable oppositions. The religious universe of the monks, with its high words, its ink and blue writing, is appreciated in the same sensual sense. Words can be sensually enjoyed—the blue veins that refer to ink return again and again as an erotic motif—and touched.

What emerges from these sensual descriptions is an embodied world that is not primarily visual but intensely sensual, through smell, taste, touch, *and* sight. Moreover, this sensuality is reciprocal. The theoretical concept that comes closest to this imagination would be Merleau-Ponty's notion of the flesh, which, in the words of Elizabeth Grosz, refers to "the condition of both seeing and being seen, of touching and being touched, and of their intermingling and possible integration, a commonness in which both subject and object participate, a single 'thing' folded back on itself" (95). Perceived as such, language will not be excluded from embodiment. Merleau-Ponty's description of language as "the very voice of the things, the waves and the forests," as "another flesh," another "wild being" (qtd. in Grosz, 1994, 102) allows us to include language and words into the sphere of sensuality, a gesture for which the groundwork has long been laid by a Christian tradition that wishes to see word as flesh, but is now subverted. Merleau-Ponty's work, read through Elizabeth Grosz eyes, offers a productive way to understand Scott's novel not as the psychoanalytic exploration of a pathological addiction to Imaginary mirroring, but as an effort to redefine the body's need to see and be seen, touch and be touched, without taking recourse to the forbidding categories of a normative patriarchal, colonial psychology. The novel's most important strategy in this aspect is the refusal to imagine the gaze as authoritative, disciplining, imperial, paternal. For Scott's characters, the gaze is almost a touch. It is sensual, curious, a tentative intermingling, as when Aelred saw his own face mirrored in that of the painted black boy, and discovered himself in turn to be the object of another gaze. This is not the authoritative, patriarchal gaze, but the gaze that strives for reciprocity, and identification.

MIMESIS AS TOUCH

At the heart of this imagination is an approach to the notion of mirroring that lies close to Walter Benjamin's rewriting of the notion of mimesis for the era of modernity. Michael Taussig (1993) explains: "Here is what is crucial in the resurgence of the mimetic faculty, namely the two-layered notion of mimesis that is involved—a copying or imitation, and a palpable, sensuous, connection between the very body of the perceiver and the perceived" (21). In contrast to a recurrent definition of mimesis as a merely iconic form of representation, then, mimesis is an "indexical [...] relation of similarity" (Marks, 2000, 138). Copy *and* contact, as Taussig repeats.

Walter Benjamin's reflection on mimesis should be placed within the context of his theorizing of the "alienation of the senses" in modernity, and the way to overcome this (Hansen, 1999, 309). Especially after the rise of fascism, people's defensive shields against stimuli would be thickened, which led to an increasing general alienation (311). Benjamin's reflection relates to Marx's theory of commodification too, which also implies an alienation from the processes of labor and human interaction—processes that are obscured in the appearance of the commodity as an autonomous entity. The notion of the contact that went into the making of the commodity has been replaced by a visual perception of the commodity. For Benjamin, modernity's visual machinery (film) was able to "restore this erased sense of contact-sensuous particularity" (Taussig, 1993, 23). Taussig likes to emphasize the relation between the emergence of a new, embodied definition of mimesis in Benjamin's work and Benjamin's acknowledgment that primitivism was an integral part of modernity. "Primitive" societies, as Taussig explains extensively, give mimesis an important role in their rituals, as an *embodied* form of copying, which may for example consist in the aim "to capture that [. . .] spirit power" of the person who is copied in a magic statuette (17). Benjamin hoped that this "primitive" quality to address and imply the body would be inherent in cinema, that might in that way reconnect to the alienated body, evade its defensive shield, and reactivate its senses (Hansen 321).

If Benjamin's theorizing occurs against the background of his profound critique of modernity, Laura Marks's study of the relation between visuality and the tactile in her book *Intercultural Cinema* is meant to open a space for the recognition of non-Western epistemologies in cinema today. Her appropriation of Benjamin's, Taussig's, and other elaborations of mimesis results in an openness to what cinema might accomplish *today* in terms of a cultural critique.

> Through mimesis we cannot only understand our world, but create a transformed relationship to it—or restore a forgotten relationship. Mimesis is an immanent way of being in the world, whereby the subject comes into being not through abstraction from the world but compassionate involvement. (Marks, 2000, 141)

Scott's novel shows that literature can accomplish the same. His writing emphasizes the spiritual element in this immanence, an aspect that Taussig foregrounded already as essential in the concept of mimesis. Emily Wroe recognizes an affinity with Wilson Harris, although the intensely *horizontal* structure of Scott's narrative prompts me to read him in a slightly different way. Scott seems less interested in delving into the womb of space to reconnect to the world than in reaching out and embracing the sensual bodies directly around him.

But the novel is not just proposing a tactile sensuality that might escape the patriarchal, imperial authoritative gaze. Rather than firmly situating it within a phenomenological tradition, I would prefer to see the novel as negotiating between psychoanalysis (and the tragedy of being excluded from the phallic symbolic order) and a potentially more liberating phenomenological sense of human relations as embodied. The second dimension does not cancel out the first, nor does it lead to a happy vision of unencumbered love and friendship. Society's forbidding patriarchal order is taken too seriously for that. In addition, the novel's acknowledgment that mimesis can be deadly resonates with the Freudian notion of mimesis as death drive, "the willingness to merge back with nature" (Marks, 2000, 143). Its insistence on embodiment cannot counter that psychoanalytical pessimism, nor does it undo the identitarian discourse of psychoanalysis. In this sense, the critique that Scott's novel leaves tensions unresolved is justified, but solving them might have only been possible by reducing the issues. What the novel has to offer is more important than a "solution" could have been: a strident critique of patriarchal societies, combined with hopeful explorations of alternative epistemologies of love that are based on dangerous forms of identification, as they may lead both to reconciliation *and* death. The many references to flowers succeed in doing both. They do not reconnect the fragmented Caribbean, as Reiss would have it. However, they do bring together a wealth of global traditions of interracial and intercultural transgression, both erotically and spiritually. As such, they reconnect dominant and marginalized histories of the world—diverse histories of violence, perversion, and intimacy.

Scott's narrative offers a thorough, innovative exploration of the interconnecting forces of binary identification (Baumann and Gringrich, 2004), mirroring, and mimesis. The author associates this mirroring form of identification with the loss of self—connectedness is obtained at the prize of the loss of self. However, he understands that loss in a less negative way than it usually is understood, as he connects it to two strong cultural codes: the religious or spiritual discourse around the loss of self in divine transcendence, and the (gay) erotic discourse on the loss of self in the sexual act. These codes are evoked by Scott's incessant references to flowers, and their complex historical codification. In addition, by defining the strategies of mirroring that organize his narrative as mimesis in Taussig's sense, Scott is able to move away from the colonial logic of the mastering, subordinating gaze, and connect to a phenomenological tradition that defines mimesis as copy *and* touch, as reciprocal visuality *and* tactility—a move made easier by the Symbolist interest in the multiplicity of sensual experience, and a Caribbean poetic tradition that values the same. Instead of refusing the (visual) patriarchal symbolic order altogether, Scott searches for, and finds, new symbolic codes in the domain of nature, that imply a relational, fragmented self, marked by a desire for merging with otherness. If this sense of self moves beyond humanism, the novel nevertheless ultimately appeals to a humanistic ethic, as Donnell (2006) points out, quoting this sentence: "(H)uman compassion can confound us" (Scott, 1998, 271). However, this humanism is strongest in the heterosexual narrator, the later, reflective Jean Marc, and one of Jean Marc's closest, gay friends. It is far less present in Aelred's passionate quest for love, which centers on the "confounding" referred to in the above quote, and the posthumanist insight that the quest for a stable identity is *not* what will save us, in the end.

NOTES

1. Reiss does signal the tension between, on the one hand, the urge to see the unifying function of the metaphor of the flowers-and-the-bees, and, on the other hand, the emphasis on diversity in this metaphor: the honey is something really other than its components, and these components are, in turn, different from each other, to such an extent that they cannot melt into one unified whole. This tension, this ambivalence, would be inherent to Caribbean literary invocations of plants.

2. A wonderfully posthuman imagination of the beehive can be found in a short story by Indian writer Vandana Singh, where we meet a man into whose ears bees are disappearing and emerging again; he explains that the beehives and computers are comparable, implicitly suggesting that the human brain also works in that way. The posthuman point here is that the human agency over the process of thought is lacking (Singh, 2004).

3. Reiss's sketch of the European literary tradition of the flower is reductive—he skips, for example, the highly subversive flowers in French Symbolist writing, from Baudelaire onward, which makes it impossible to situate all subversiveness in the Caribbean tradition only.

4. Another comparison between Scott and Dabydeen is that both read a slave boy's (Mungo/Jordan) story as not just on racial, but also sexual subjugation. Dabydeen focuses on the boy Mungo, that was held captive by the captain of *Zong*, the slaveship pictured by Turner on his famous painting *Slave Ship (Slavers Throwing Overboard the Dead and Dying, Typhoon Coming On)* 1840, discussed in Emery. Emily Wroe also points at similarities between Dabydeen and Scott.

5. Homi Bhabha's analysis of the visual dynamics of the colonial condition, which builds on Fanon, also emphasizes the reciprocity and instability of the processes of identification involved (Bhabha, 1994, 40–65).

6. One might object that Benedict's homosexuality is the reason for his inclusion in the chain. I would nevertheless—inspired by Fanon—prefer to see *desire* as the link, not (marginalized, oppressed) identity; not necessarily sexual desire, but a desire for intimacy with other men, of whatever kind and intensity—a desire that is violently repressed in the construction of heterosexual masculinity.

7. The fact that the compassionate brother, who at first does not want to know everything about the brother he had been ashamed of, finally comes to terms with his brother does not necessarily mean that this novel has a more positive view of the possibilities to relate to the suffering, gay brother's predicament than Kincaid's novel (discussed in chapter 5). This surviving sibling is as fully aware of the risks of identification as Kincaid's narrator, but, because of his gender and his belated entry on the scene (safely after his brother's death), isn't as vulnerable as his counterpart in Kincaid's text, and is therefore able to drop his defenses.

8. Shani Mootoo's novel *Cereus Blooms at Night* offers a more radical exploration of that possibility (chapter 2).

9. Patrick Murphy's (2012) gender critique of the concept of the sublime would also be relevant here.

10. In her important discussion of the Creole epistemologies that bear on ecology and sexuality, literary scholar Natasha Tinsley (2005) shows how, in specific love songs in the Dutch Caribbean, references to flowers serve to imagine the alternative identities of (Creole) women who refuse the "hegemonic model of heterosexual womanhood" (180). Tinsley's observations, together with the association made by Hammer (in Goody, see above) between flowers and nonhegemonic sexual practices among women, suggest the need for more transcultural research of women's tradition of imagining female sexuality through images of flowers.

11. It is illuminating to see that Scott's flowers also differ from Mootoo's, as the latter refer to the hidden horrors of colonialism, and are therefore more closely associated with trauma (as in Kincaid's work). Scott's flowers are more benevolent, in that sense, than Mootoo's. But both writers refer to flowers as alternative models for transgressive identities. Strikingly, Scott, as a male writer, appears to have many more cultural codes at his disposal to refer to such (gay) identities than Mootoo; the cultural codes that signify transgenderism, transvestism, or lesbianism are much less available than their masculine counterparts, though Caribbean literature offers several strong images, such as the metaphor of the insects and butterflies, both in Mootoo and Sarduy (1989) (see chapter 7). Perhaps this is why Mootoo's flowers stay closer to an understanding of nature as "other."

Chapter Four

Are We All Connected?

A Critique of Empathy

Much lauded writer Jamaica Kincaid made her name with some highly evocative prose texts that were poised between the Caribbean postmodernist and the Caribbean modernist, shifting into the registers of postcolonialism, though not without subverting them, all while delving deep into Caribbean epistemologies (Hoving, 2001a, 184–237). In the early 1990s, Kincaid started publishing garden columns in *The New Yorker*, in which she pondered the colonial violence inherent to gardening ("Flowers of Evil"). In 1999, after having edited a collection of essays and poems about plants (1998), she published *My Garden (Book):*, but whoever buys *My Garden (Book):* for gardening tips will be surprised. In these pages she writes about the connection between colonialism and gardening, about the criminal germ of possession that leads to the abduction of plants, about the destructive desire to know plants by (re)naming them, and about colonial and anticolonial ways of gardening and landscaping. For Kincaid, gardens inevitably refer back to colonialism, as most plants are not native to the garden where they are growing. European botanists have brought exotic species to European gardens, in the course of expeditions that were closely related to colonizing ventures.[1] In a thought-provoking ecocritical essay on the lack of cultural diversity in the emerging field of ecocriticism, Canadian critic Susie O'Brien discusses Jamaica Kincaid's garden stories to bring out Kincaid's sharp perception of the complex relation between culture and nature (2002). Her essay is an eloquent example of the way in which Caribbean writing can productively intervene in mainstream ecocriticism. O'Brien shows how Kincaid's work problematizes certain naïve assumptions about the potentials of ecocriticism, mainly the assumption that ecocriticism would somehow endow the humanities with a new social relevance, and thus effectuate a return to the "real world" (169). This desire to go "back to the world" is supported by arguments, that, as O'Brien warns, risk "a kind of conceptual sloppiness" (169)—for mainstream ecocritics all too easily assume a direct link between their own critical reading of "the world in the text," and the possibility "to *change* the world" (169). This idealistic opinion is "sanctioned by the ecological cliché that 'everything is connected to everything else'" (169).

Kincaid's work, by contrast, shows us the problems in the desire for an unmediated return to nature, as "the physical world described by Kincaid is difficult to access" (170). One reason for this difficult access is that a *local* environment has come into being by transnational "processes of cultural and environmental dislocations" (170), such as the (colonial and neocolonial) transnational importation of plants, the colonial botanical expeditions, the import of colonial aesthetics, and so forth. Kincaid's focus on gardening makes it clear that nature is shaped by culture, and vice versa, but it also teaches us, and O'Brien emphasises this point: "to resist the urge to make them (nature and culture, IH) analogous" (177). While nature and culture are intertwined, Kincaid also insists on their incommensurability (179).

A few years later, O'Brien's plea for a more culturally diverse ecocriticism is answered by an important edited volume on Caribbean literature and the environment (DeLoughrey, Gosson, and Handley, 2005). DeLoughrey et al. emphasize that Caribbean writing is not merely referring to nature to bridge and connect, with the aim of creating a noncolonial sense of place and self (see Reiss in chapter 3, earlier). It is also disconnecting, subverting, and emphasizing opacity, incoherence and incommensurability. DeLoughrey et al. point out that the recuperative imagination of the Caribbean as continuous (for example, as African) "has led to facile attempts to smooth over the inherent discontinuities of New World history" (20). Paraphrasing Glissant, they state that "every step forward in forging a new identity and sense of place from the fragments created by New World experience means leaving behind an imagined whole [. . .] historical reconstructions [. . .] may largely be an act of imagination or desire for a wholeness that is not achievable" (21). How then, do the editors ask, "does a writer achieve a sense of place?" (21). This is also an ecocritical question.

In the next two chapters, I want to take this question seriously by approaching Kincaid's text as a literary project motivated by the problematic condition of the Caribbean (and the globalizing world). I will carefully explore the unusual ways in which the text manipulates connecting and disconnecting devices to obtain its aims. In spite of the prominence of these connecting strategies, I will argue that Kincaid's relation to her fragmented human and nonhuman environment is not energized by a desire for connectedness, as she sees connectedness as potentially destructive, but is instead motivated by her effort to come to terms with life's materiality. Her unusual literary strategies of connecting and disconnecting serve that last writerly aim. Here also lies the answer to the question why references to nature play such an important part in her work: they allow her to relate to nondiscursive matter.

After a short consideration of some discussions around the organicist concepts of coherence and connectedness, I will continue my argument by responding to a certain form of reading of Kincaid's novel *My Brother* that is organized around the concept of empathy—the ability to emotionally connect to others. This allows me to offer another reading, and outline a different tradition of environmental writing than we have so far encountered in the chapters above, and that was concerned with the need to heal the colonial fragmentation and dividedness through reconnection (which mainstream ecocriticism would see as the most obvious strategy). This alternative tradition is marked by a strong desire to return to stark matter, in spite of the paradoxes to which Susie O´Brien draws attention.

My discussion of empathy, connectedness, and coherence will help to adopt a critical distance to some key concepts in organicist thought, which is a way of thinking that has shaped approaches to nature for a long time. In this way, I hope to contribute to the postorganicist turn in ecocriticism that is so eminently visible in Caribbean literature and criticism. On the other hand, I do not want to argue that Kincaid's writing would unambiguously tie in with the posthumanism that counters organicism, or new materialism. However, a materialist reading does help to get a clear grasp of the unusual devices of connecting and disconnecting through which she creates a sense of place. We cannot understand these without an insight into the psychological dimension of Kincaid's writing. In this and the next chapter I will situate Kincaid's writing on the fragmented spaces of the Caribbean in relation to organicism and post-organicism, humanism and posthumanism.

CONNECTEDNESS, COHERENCE: SOME MADDENING ASPECTS OF ORGANICISM

In spite of their centrality to ecocriticism, organicist concepts have been subject to severe critique, especially since the emergence of postmodernism. The concept of connectedness, for example, served as the butt for a provocative critique by Halberstam and Livingston (1995), who argue that "(t)he organicist notion of connectedness—and its most extreme mystification, the Romantic imagination—had been invented as internalizations and depoliticizations of dominant material interests and their power/knowledge grid" (2). One might indeed support the diagnosis that concepts that imply connectedness (i.e., shared humanity, identification, analogy) effectively obscure unequal social relations. This political critique is only one of the many attacks on organicist notions such as connectedness and coherence. Timothy Morton (2007) offers a wonderfully thought-provoking critique of organicism in *Ecology without Nature*, in which he, however, also argues that the wish to radically break with Romanticism would be counterproductive. There are inherently contradictory aspects in this tradition, such as the acknowledgment that the effort of unification is futile. "Environmentalist values (complexity is good, the world cannot be totalized though it is a whole)." Morton identifies many of these as "slices of Romantic organicism" (97), which can also be productive. His remarks show the broad scope of organicism, as it would even inform insights that are often seen as *critical* of organicism: Glissant's work, for example, circles around the desire to offer a poetics of the world as a whole that is nevertheless not totalizing, as we will see in later chapters.[2]

It seems undesirable to just radically discard any form of connectedness, even if the concept is inherently contradictory and often problematic. One reason is that it is difficult to imagine what access to knowledge would be offered in writing that radically *refuses* symbols, metaphors, and therefore coherence.[3] According to Victorianist Christopher Herbert (2004), the idea of coherence is "(t)he guiding principle of specifically modern intellectual enterprise" (185): "Understanding it (an object of study, IH) means nothing other [. . .] than describing the various ways in which it manifests its twofold natural endowment of coherence, of adhesiveness or stickiness" (185). This notion is so much naturalized that "(t)he incoherent, if definable at all, is that which makes no sense" (185).

By the mid-nineteenth century, the notion that a scientific approach consists in defining the "manner of participation (of things, IH) in systems of coherence that [. . .] are sure to be complex and ambiguous" (186) was closely affiliated to the new mode of thought that most notably emerged in Darwin's evolutionary theory. This principle of relativity held that science should be concerned with relations, not with the properties of seemingly autonomous, individual objects or events (187). Herbert describes how the notions of coherence and relativity were then developed in biology and sociology (Herbert Spencer), linguistics (Ferdinand de Saussure), and ethnography (Franz Boas, Ruth Benedict). Remarkably, when the principle of coherence finally made its way into literary theory (notably New Criticism) after being "fully elaborated long since in every field of investigation that had embraced a modern style of thought," it was received as a novelty—even if, as Herbert explains, it made its impact "seventy-five years or so late" (196). There was no acknowledgment of the paradoxes and problems inherent to the notion of coherence. The view that everything depends on everything else appeared to result in a dizzying understanding of the world as virtually endless, and therefore impossible to know. In short, says Herbert, "coherence and incoherence are seemingly reversible categories" (204). The literary scholars who adopted the principle of coherence in the late 1930s nevertheless insisted that this approach would lead to the certain and objective establishment of a literary work's "total intention" (197). It is only far into the second half of the twentieth century that postmodern scholars would begin to criticize the concepts of coherence and unity and adopt new concepts that express the partiality, situatedness, and incoherence of their objects of inquiry. Herbert's illuminating consideration of coherence is a valuable contribution to a critique of the concept of connectedness, and, together with the critical reflection offered by scholars such as Halberstam, Livingston, and Morton, invites a critical scrutiny of the concept in ecocriticism. I will now engage in an analysis of the way in which the notion of connectedness functions in the assessment of Kincaid's literary strategies—where it is sometimes translated into a prescriptive psychological and moral value, that is, empathy.

THE RISKS OF EMPATHY

In her discussion of the politics of sexuality in Caribbean writing, Alison Donnell (2006) describes the mirroring aesthetic in the novels by Mootoo and Scott as a "model of compassion and openness" (214). This "radical possibility of symmetrical reflection" (213) is based on the capacity for empathy: "recognitions of self as other are affirmations of the significance that empathic identification may hold in terms of transgressing the boundaries policed by national and sexual identity politics" (213).

In stark contrast to her praise for these authors, Donnell takes Jamaica Kincaid to task for falling short of this demand for empathy in her novel *My Brother*: "This absence of empathy and of identification is marked in the text by a repeated assertion of her difference via her family [. . .] and [. . .] in its naturalisation of heterosexuality" (Donnell 2005, 225). The brother in the novel, a closeted homo- or bisexual man who is dying of AIDS, is indeed depicted as an irresponsible, unarticulated person, until, in the course of his illness, he is represented as a horrifying wreck. The narrator in this seemingly autobiographical account offers important help, but refuses to take him into her "hard-earned order of my life" in the United States, where he might have had better medical treatment (Kincaid, 49). Donnell is not the only critic to feel uncomfortable with this account. Greg Thomas, for example, is at least as harshly critical, and so were a number of other critics amidst the more general praise of the novel (Thomas 2007, 121–25; see a reference to other critical responses in J. Edwards, 2007, 10; Smith and Beumel, 2006).[4]

It is hard to disagree with Donnell's praise of empathy as a healing response to the destructive strategies of othering in Caribbean society. An inclusive attitude might indeed counter practices of exclusion. But what is empathy exactly? And is it really empathy that Kincaid's narrator is missing? Assuming that empathy is a form of identification, as Alison Donnell suggests, I want to take a closer look at the narratives of difference, sameness and (dis)connectedness that are expressed in Kincaid's novel. *My Brother* is, apart from all else, a continuous play of comparisons, contrasts, elaborations, and qualifications.[5] At any moment, the narrator will interrupt her argument to comment on an object, event or attitude, qualify it, contextualize it, and relate it to another object, event or attitude, in a highly associative way. Often, such elaborations will occur in a repetitive style, in symmetrical sentences.

> At that moment [. . .] my mother's life (a life she might have had in mind, or a life that had become a nightmare, how could I, how can I know) collapsed [. . .]. Her husband (the man who was not really my own father, my brother's real father) was old and sickly. (128)

> but the frozen waters of Lake Michigan were a blue not like a blue my brother had ever seen, it was not like the blue of the Atlantic Ocean in the West Indies [. . .] nor was it like the blue of the Caribbean Sea. (166)

In the first quote, the narrator takes her distance from a possible qualification of her mother's life, though she does qualify the nature of her knowledge about its collapse. She then goes on to specify her own relationship to her mother's husband, which (again) dissociates her from him. These short interventions question the transparency of the narrator's family, her natural relations to her parents, and create distance between her and her family. The associative strategy that is especially apparent in the second quote might be used to create a signifying network, in which the color blue functions as the thread that links the narrator with her brother. However, at the very moment this meaning might emerge, it is questioned and dismissed. After the admirable associative play with the color blue throughout this part of the story, which returns in the clothes the narrator buys for her brother's funeral, the airplane that carries her to the funeral, and so forth, the narrator comments: "But the color blue did not run through all my memories [. . .]; on the whole, every scene, every memory remained itself, just itself" (170).[6] At many different moments, the narrator muses on the relevance of the use of associations and repetitions, through a recurring motive: "If it crossed his mind that this tree, coming out of a dormancy, a natural sleep, a temporary death, or just half-dead, bore any resemblance to him right then and there, he did not say, he did not let me know in any way" (80). It is significant that the play with associations very often leads to animal and plant life, often in the state of dying or killing. More and more, the narrator admits that she cannot find comparisons or meanings: "his heart beating like something, beating like something, but what, but what, there was no metaphor" (107). Sometimes she just observes "that all these events in my life would come together" (137), without offering any deeper signification.

There is a good reason for the refusal of these strategies of connecting. In the case of the color blue, the narrator explains that "sometimes a certain color might make a memory more vivid and sometimes again, not so at all [. . .] but the absence of color does not mean an absence of truth" (170). The statement becomes understandable when one reads the exclamation below: "Oh, how I wished never to think about him again, never to see blue and say, 'What did Devon think of that, such a color'" (170). The association between the color and her brother's death causes her a pain that is perhaps more vivid than if she were to have thought of her brother without this association. Nevertheless, the truth, the value of that memory, is not dependent on its intensity. Neither vividness nor intensity of feeling are indications of a hidden, true meaning. And truth is independent of its expression through a symptom. On the one hand the narrator proposes different well-known stylistic strategies to create literary connections, and weave a coherent text (recurring motifs, metaphors), but on the other hand, she disturbs them.

There are different ways to read this critique of signification. To begin with, a psychoanalytical approach seems to suggest itself: here, Kincaid acknowledges the workings of Freudian displacement, where a certain emotional intensity (affect) is disengaged from a certain element in a dream, and then cathected on another, unrelated object, and then, possibly, on a third or fourth object, and so forth. The narrator accepts that the intensity of a certain experience may be insignificant, as it relates to an original experience that is now displaced, but she also opposes the (both psychoanalytical and philosophical) idea that the significant realities and truths in one's life would be accessible through language and metaphor.

I will come back to the latter insight, when I discuss the epistemological dimensions in this text, but I will first venture a loosely interdisciplinary reading here, invoking postcolonial theory, psychoanalysis, phenomenology, and Caribbean trauma theory, to argue that Kincaid's resistance to signification is linked to her wish to counter the (supposedly enabling) connecting strategies of empathy and sympathy. While this psychoanalytical discussion leads away from environmental debates on connectedness, it will be helpful for articulating a well-founded critique of the notions of connectedness and empathy that lie at the heart of ecocriticism.

In the first place, then, Kincaid's distrust of signification can be related to the pathological condition of the (post)colonial, which is, in her work, often exemplified by the narrator's mother. As the narrator has explained elsewhere, the postcolonial Antiguan condition is characterized by an inability to acknowledge temporality and causality.[7] If Antiguans are unable to discern past causes for present-day conditions, the narrator's mother even forbids her children to remember (6, 25, 75). She aims at complete (idiopathic) identification and refuses her children an autonomous sexuality. She even throws stones at her oldest son when he has found a partner (188). The narrator suggests that the mother forces her children to remain with her within an Imaginary order, in which they will keep being absorbed by her presence, and are eternally dependent on her. And she succeeds. Her sons never set up a household of their own, staying with their mother instead (53). The narrator states that she would probably have shared her brother's destiny, if she had stayed (75, 90). The destructive maternal preference for mirroring and sameness might go some way to explain the daughter's abhorrence for this aesthetic. Though the narrator always begins by adopting these signifying strategies, she also opposes the infinite forms of semiosis that are related to both the Imaginary and the dream work.

It is characteristic of Kincaid's work to establish a link between the Imaginary predicament and the postcolonial condition of her birthplace. An obvious reason for that link lies in the fact that postcoloniality, in her eyes, is largely a continuation of the colonial condition, which is marked by the dynamics of the Imaginary (see Bhabha's accounts of mimicry, JanMohamed). But Kincaid, as most other postmodern and postcolonial writers, is interested in the role of premodern epistemologies too. In certain senses, the premodern and the postmodern are affiliated.[8] The Antiguans' dismissal of causality echoes the modern crisis of causality when, in the nineteenth century, the organicist concept of coherence began to gain popularity (Herbert, 192). Playing on the affinities between the modernist interest in primitivism and her own assessment of (post)colonial Antiguan popular culture,[9] Kincaid highlights the destructive dimensions of this position. She suggests that the Antiguan/maternal attitude to identity is close to certain so-called "primitive" notions of identity, in which the self threatens to be erased. Her approach echoes the analysis of the "primitive" sources of mimesis as a magical strategy of heteropathic identification offered by Michael Taussig (see chapter 3), or the collapse of sympathy and identification in primitive societies referred to in the work of phenomenologist Max Scheler, at the beginning of the last century.

Scheler (1992) discussed the complex notion of identification, first by differentiating the two opposite kinds, idiopathic and heteropathic identification, then by offering a number of instances of both. His first example of heteropathic identification is the process "of identification in *primitive* thought [. . .] for instance, the identification of each member of a totem with an individual member of the totem species," or with their ancestors (60). Scheler emphasizes that this is not merely the case of being "like the ancestor, or guided and ruled by him, but actually is, in his present life, at the same time one of his ancestors" (60). In this dynamics, difference is erased; it implies the erasure of the self. This condition has to be differentiated from an ancestor cult, which, in his words (and Kincaid could have used them), "already represents a first stage of *liberation* from the primitive identification" as it "presupposes a recognition that the two are *distinct* individuals" (60).

One might wish to problematize Kincaid's intertwining of postcolonial trauma, the Imaginary, and the primitive. More important, however, is what emerges from this analysis: the acknowledgement that strategies of connecting are building on older ("original habits of," as Scheler 60, has it) strategies of identification, and thus on the *erasure* of difference—and that these strategies may always collapse unto their "original" forms, in which the self is erased. Both Kincaid's and Scheler's work suggest that one has to *work* to move away from the cannibalistic, complete identification at the heart of older rituals, and establish a more distanced position, for example in the form of sympathy or empathy. With the rise of organicism in the nineteenth century, which emphasizes the world's boundless interconnectedness, the pull of identification is not at all diminished. The more "primitive" form can emerge again at any time, as Scheler argues, for example in the mass cult of the Leader. The threat of the erasure of self is still very present in all, also the more modest, forms of identification.

In contrast to Alison Donnell's evaluation, then, I would suggest that it is rather an *excess* of empathy than a lack of empathy that forces the narrator to take her distance from her family: "I am so vulnerable to my family's needs [. . .] that from time to time I remove myself from them" (Kincaid, 20; see also, 117). The narrator fears her own death when she goes to visit her dying brother (Kincaid, 92). She even uses the notion of intimacy in bleak associative chain that leads us back to the relation between humans and the nonhuman world: "to be so intimately acquainted with the organism that is the HIV virus is to be acquainted with death" (95). Intimacy is knowledge, and knowledge is deadly. This is where one's intimacy with a cannibalistic mother would lead. Intimacy holds the danger of contagion; see, for example, page 173. Empathy is predicated on identification, and identification (with the stronger maternal presence, here) amounts to a lethal loss of self. As literary scholar Ann Astell argues, following René Girard, "'[e]mpathy' and 'contagion' are semantically related" (Astell, 2004, 116). Max Scheler makes the same point: "The true *sense of emotional unity*, the act of identifying one's own self with that of another, is only a heightened form [. . .] of infection" (59).[10] The security offered by class can furnish a defense against this infection. Kincaid's narrator analyzes empathy as a class-related privilege one cannot always afford; one can only pity others when one is well off oneself (125). Within the context of her family, though, intimacy and empathy would lead to her annihilation. In addition to defining pity as a privilege, then, the narrator also describes it as an attitude linked to (protecting) superiority (186).

I have shown why Kincaid's text rejects empathy and identification as a primary strategy to forge a relation (to the Caribbean, to the world) that would make sense of its narrator's surroundings. Let us now consider the strategies the text adopts instead. Above, I mentioned a historical strategy, used in *My Garden (Book):*—namely, the analysis of every environmental location (including the Caribbean) as a crossroads of transnational economic and ecological dislocations. In *My Brother*, I discern a few other strategic moves. The most important one is her narrator's decision to refrain from judging (in contrast to the highly judgmental mother, 148), and take her distance instead. But this is only one possible step; a second strategy, equally productive, is the acknowledgment that she does not know her mother at all (93).[11] Here we find the paradox that is elaborated in Kincaid's writing. On the one hand, knowledge (understood as predicated on identification, intimacy) can be lethal; on the other hand, she cannot refrain from an attempt to know. Her writing, which she describes as an effort to make sense of what is happening, amounts to a survival strategy in which she has to navigate between approaching her subject (and risking destruction) and taking her distance (see 196). Kincaid's writing is an endless effort at bringing together events, juxtaposing, relating, and contrasting them, but staying on the outside by keeping ambivalence intact, maintaining distance, suspending judgment, and always reflecting on possible meanings, while often refusing these meanings at all. By connecting and disconnecting in the same gesture, she is able to face the world without being devoured.

Far from being productive means of connecting, empathy and sympathy may hamper one's relation to the world, and revive the trauma of being erased. However, there is another aspect to empathy that needs to be addressed: empathy can also hamper knowledge. The following sections are dedicated to a discussion of Kincaid's work, not as a psychological, but an epistemological project. It is here that the environmental aspects of her work are most visible. Does her work represent a critique of earlier ways of knowing? Does she offer another way of knowing—and especially another way of knowing the postcolonial, traumatized, transnational space she inhabits? What is the role of connectedness in these strategies of knowing? How can her literary struggle be relevant to ecocriticism?

EMPATHY AS AN OBSTACLE TO KNOWLEDGE

Max Scheler's work suggests that empathy may well be seen as an obstacle to knowledge, a point repeated in the 1940s: "Scheler's precise formulations drew our attention to the important fact that the more sympathy is a result of an identification, the more untrue it becomes" (Steinfeld, 1944, 120–21). According to literary scholar Deborah Nelson (2006), a comparable statement is defended in Hannah Arendt's *On Revolution* (1963), where she "carefully dissected the 'devastating' effects of compassion in political life" (87). Arendt "imagined the consolations for pain in intimacy, empathy and solidarity as *anaesthetic*" (88). Nelson opposes two groups of ethical models of relation: the dominant models "since the tragedies of the mid-century have tried to bring the self face-to-face with the Other." As such, they can be contrasted to

> a countertradition of ethical relation, one that seeks not to come face-to-face with the Other but to come face-to-face with reality in the presence of others. Since reality and the Other cannot be faced at the same time, [Arendt et al.] chose to face reality, however psychically wounding. (88) [12]

Kincaid's position is of course very different. For her, empathy is hardly anaesthetic, it is self-destructive. The reality she wants to face is twofold, consisting of (post)colonial traumas that are closely linked to her primary other, her mother, and of the harsh, nondiscursive materiality of life itself. Nevertheless, like Arendt, she also accepts the ordinariness of pain (Nelson 88) and the importance of factual truth, which would often be concealed by so-called positive emotions (Nelson 92). In Kincaid's texts, knowledge can therefore not be obtained by introspection, as this would only lead to a host of contradictory feelings that give no insight in truth. Nor will introspection necessarily lead to the healing of trauma (which is in accordance with the theory that assumes that such a healing can only occur through reconnection with others; see Herman, 1994). At a certain moment Kincaid's narrator ponders her own inner "feelings I might not ever understand, feelings that everyone who knows me understands with an understanding that I will never know, or that someone who has never met me at all would understand as if they had made up my feelings themselves" (109). Knowledge about an inner reality is unavailable to those who embody that reality. It is only from a distance, from the outside, that something can be known. Empathy does not seem decisive here: people might understand the narrator's feelings *not* as if these were their own, but as if they had *produced* them for her (compare the Wittgensteinian analysis, where empathy consists in an active simulation of otherness). Thus, even incomprehensible private feelings are approached as external products.

It is only *from the outside* that the narrator will be able to relate to what happened to her brother. Empathy might lead to her own destruction, while not yielding any real insight in her brother's mind. His own feelings and desires might not even have been clear to himself. Assuming that one can understand others by reading their thoughts is a misunderstanding: human beings are not transparent. Identification will not help, nor will any clear analogical form of reasoning, as no analogy will reveal the ungraspable, contradictory complexities of someone's life. As the narrator argues, people do not have just one identity (Kincaid, 171). Writing is therefore only possible after dissociating one from those who are like her and who know her (162). This statement is quite at odds with identity politics, but also with the organicist view that sees identities as inextricably bound up with their surroundings, and that holds that they can be fully known from within. But it does tie in with a broader Caribbean ecocritical view that breaks with organicism. According to literary scholar Jana Evans Braziel, the early Glissant outlined that the damaged Caribbean cannot be grasped directly, from within. A detour through diaspora is needed to obtain that knowledge. Braziel quotes him with approval: "distancings are necessary to Relation" (118). This alternative view often testifies a close affinity to (new) materialism and posthumanism.

Kincaid's decision to take distance from her family, her refusal of empathy, identification, and analogy, then, may be explained as a dismissal of the hermeneutical position. She distrusts empathy, introspection, or "Verstehen," as methods to obtain knowledge. She, too, sees distance and detours as inevitable preconditions to knowledge. Kincaid's double strategy to relate to her cannibalistic family (taking distance, claiming complete ignorance about them) implies a two-fold critique of an organicist approach that understands the world as holistic interconnectedness, that can be understood from within. In the next section, I will discuss the merits of a possibly more modest means of connecting—through analogy—and the role it plays in Kincaid's work.

POSTMODERN ANALOGIES, AND KINCAID'S POSTHUMANIST TURN

Kincaid's work is not interested in evoking coherence to create a sense of place. Nevertheless, it makes ample use of analogies, even while it shows a certain distrust of analogy as a means to produce knowledge. Such a distrust might not only be explained by the fact that analogy may obscure relations, but, more basically, because it would be *in principle* unable to produce real knowledge. Nevertheless, Kincaid's critical, postmodern, playful use of analogy, in which she links human and nonhuman life and death, is highly effective, especially as an alternative to the creation of coherence.

Analogy, harking back to the Middle Ages, has a long history in Western epistemology, in the Foucauldian sense of the term. In the centuries that followed its emergence it became synonymous with equivalence and equality. We recognize its presence in the Renaissance episteme of sameness and resemblance, and see how the classical emphasis on universal, spatial categories informed the translation of analogy into law. In the seventeenth century, Newton reinforced this emphasis on the resemblance and identity that were part of the concept of analogy, by assuming that the analogous qualities discovered during his experiments were universal. In this way, analogy became a form of translation—a translation from the known to the unknown, from the particular to the universal, or, as literary scholar Eve Tavor Bannet (1997) puts it: "analogy 'maps' elements of a 'source domain' onto a 'target domain'" (657). The known is thus projected on the unknown.

The critique of analogy as a means to obtain knowledge took many forms,[13] and it is not restricted to postmodern philosophy. It has been uttered by many philosophers and semioticians.[14] Followers of Wittgenstein, as testified in a study by philosophers Kögler and Stueber (2000), also doubt the soundness of the method of analogical inference. Kögler and Stueber focus their critique on the moment of mapping, which would be a movement from the known to the unknown (as Bannett has it). Contrary to those who contend that the empathic introspection occurs by taking oneself as a model for the other person's thoughts and motivations, Wittgensteinian thinkers argue that one merely tries to *simulate* the other person, to imagine *their* inner life—so that there is no inference from (a known) "I" to (an unknown) "other," but merely between the "other-simulated by-the-self" and "the other" (Kögler and Stueber, 9–10). Kincaid would add that one does not know oneself anyway, so that a comparison between self and other does not lead to any superior knowledge about the other.[15]

One of Derrida's critical responses, Bannett (1997) explains, was "a parodic repetition of the kind of analogical reasoning that it also deconstructs" (661), through his unlimited play with resemblances, which "turns equivalence into equivocation or undecidability" (662). Whereas Derrida thus deconstructs the assumptions of the law-governed relations between self and other, Wittgenstein shows that language is not governed by fixed, mechanical laws or rules, but should rather be considered as a game, which allows us to change and subvert rules, or invent new rules (663–65). The analogical game can therefore also be used in new ways, acknowledging difference and resemblance, building bridges in spite of, and fully aware of, incommensurability. Thus, it would open up the new possibilities of language to connect events to their possible contexts. Language is less concerned with truth than with the creation of possibilities (669).

It is this latter approach that seems especially helpful to understand the workings of analogy in Kincaid's texts. Kincaid's writing is an experiment with the powerful game of opening possible connections, departing from the explicit assumption that analogies are never simply there, but only created. Moreover, she stresses the fact that her analogies do not move from known to unknown, but from unknown to unknown. At the same time, she is anguished by the meanings that are thus evoked. The analogous game with the color blue leads her into painful political entanglements, into the networks of colonialism, exoticism, tourism, migration, funerals, death—until she realizes that these varying ways of framing are not necessarily concerned with truth at all. If traditional analogy, as an inclusive practice, might work against the logic of abjection because it emphasizes sameness (note 12; see also Kristeva qtd. in Paquet, 2002, 244–45), Kincaid's postmodern analogies do often extend to the abject (brother, mother, Antigua), while emphasizing both cannibalistic closeness and horrifying difference, followed by a sudden distancing—and this shows Kincaid's persistent unease with both traditional and postmodern forms of analogy. We can certainly explain this unease from her diagnosis of the game of analogy, resemblance, connecting, and copying as rooted in the destructive, traumatized, maternal Imaginary. Even a self-reflective, playful use of analogy, that denies the stability of the laws that would govern identity and difference, cannot resist that pull.

The most important point is that Kincaid keeps coming back to the question of truth. She is interested in the hard, nondiscursive, tangible truth that is lodged in or below the play of signifiers. If she stresses that she moves from unknown to unknown, she also articulates her desire for a material foundation of knowledge. In this aspect, her work is quite at odds with the postmodern view of deconstruction as a critical strategy. Kincaid's work may be too deeply marked by the violence of the maternal Imaginary that also characterizes the (post)colonial condition to be able to hope for the redeeming qualities of deconstruction. The only way out seems a retreat from the Imaginary—that is, a return to the Real.

Here we can begin to understand the significance of Kincaid's creation of analogies between the human and the natural world. Braziel quotes one example of the "parallels between human and natural degradation, between genocide and ecocide" in Kincaid's work: "The Carib people had been defeated and then exterminated, thrown away like the weeds in a garden" (1996, 17, qtd. in Braziel, 2009, 114). This equation comes as no surprise for those who know Kincaid's analysis of the intertwining of colonialism and botanical expeditions in *My Garden (Book):*, or to those who read Kincaid's memory of how "we" (the Antiguans, the Caribbeans), during slavery, "were capital," like natural products such as sugar and cotton (1988, 37–38).[16]

Kincaid's work with analogy does not serve to create a sense of coherence in the worlds of things and humans, but it allows her to rethink the human within the frame of materiality. I have read her fascination with materiality as a psychological strategy of survival, but it also ties in with a broader epistemological shift. After the nineteenth-century emphasis on genealogy and continuity, and the rise of organicist notions of connectedness, in the twentieth century a critique emerges on the tendency of the concepts of identity and analogy to erase difference. The postmodern acknowledgment of the existence of otherness *in* the self (not as opposed to or continuous with the self, as analogous reasoning would have it) opened up a fascination with otherness, death, finitude, and the body. Like Glissant, Kincaid finds that irreducible material otherness in the acknowledgment of the *opacity* of the world's locations. By acknowledging that opacity, which is the result of the many social, economical, environmental, cultural, and political interactions that have shaped that location's untranslatable specificity, a postorganicist understanding of connectedness emerges. For Kincaid, the opacity of place is also connected to its physical materiality that, in spite of its being saturated with violent histories, also resists appropriation by social or political discourses.

Above, I suggested that this materialist reading makes it irrelevant to understand Kincaid's writing as a search for a new identity, or a new discourse.[17] It invites us to understand Kincaid's evocation of plants and trees as something other than metaphors that would help to bring out some aspect of the human condition. The relevance of Kincaid's turn to the imagination of plants and animals lies in the fact that it leads her away from hermeneutical or identitarian approaches, into a more materialist, posthumanist, postorganicist approach. At the same time, however, this material turn is always psychologically motivated.

FROM A PSYCHOANALYTICAL READING TO A DELEUZIAN/CREOLE READING

This reading opens a new debate. For one, the notion of opacity is not unproblematic. How should we read Kincaid's embrace of the idea, if we want to see it as more than the psychoanalytical desire for the Real? Glissant's notion of opacity is closely affiliated to the much-debated concept of singularity. I will address this open-ended debate in chapters 5, 6, and 10. For now, I want to refer to the moment in Natalie Melas's discussion of different understandings of the concept of singularity, to bring out an important aspect of Kincaid's investment in materiality and opacity—namely, that for her, it opens less a mode of knowing than a mode of experiencing.

In a debate about postcolonial, critical imaginations of the community, Melas discusses Jean-Luc Nancy's view of the individual as a *singular* being. This view resists both an autonomous approach to the individual, and an immanent model of the community that would erase differences and result in fusion (2007, 97). Melas, who supports Nancy's approach, clarifies it by opposing it with the "Hegelian notion of mutual recognition as the dialectical model of communal identities" (98)—a notion which I extensively criticized above, in the debate on Kincaid's resistance to empathy. "Singularities," Melas explains, paraphrasing Nancy, "are not the recognition of the same, in fact the singular properly knows or apprehends nothing, but rather '*experiences* her similar'" (Nancy, 82, emphasis NM, qtd. in Melas, 99). "No longer the recognition of sameness nor the sublation of difference, similarity becomes differential, an effect of externalizing difference" (99). Melas writes that Glissant and Nancy agree that "there is nothing recognizable in the other, which finds its analogue in Glissant's notion of 'consensual opacities'" (105). Glissant, however, relates the density of the other, or the denial of that density, to their relation to place (106).

There are two important insights here. First, in Nancy's perception, one does not *know* the other, but *experiences* them (as similar, or different). This insight brings me back to my recurrent observation that Kincaid, in her writing, tries to evade knowledge as much as produce it. Though my discussion of Kincaid's strategies of knowing has focused on issues of representation above, I would now like to emphasize the important element of affect. Second, in Kincaid's work, the experience of the other is intertwined with one's relation to place, be it as the owner of private property or as dispossessed. That is, one's experience does not spring from a classical phenomenological introspective perspective, but from an awareness of one's spatiality, that is, from the outside. This, then, is the moment to return to my original analysis of Kincaid's writing not as an effort to represent a traumatic truth, but as a performative, affective effort to relate to a painful experience—*from the outside*.

Claire Colebrook (2001) helps us to understand this device with the help of the Deleuzian concept affect. She explains that affect should be differentiated from affections—affections "are what happens to us," (disgust, or the recoil of the nostrils at the smell of cheese)" (21). Affect is the same force, but then seen as disengaged from the body that experiences it: "[a]ffects are sensible experiences in their *singularity*, liberated from organising systems of representation" (22). Kincaid's literary commitment to affect and singularity invites a materialist reading.

At first sight, my initial psychologically motivated reading of Kincaid's work as the effort to undo the disconnecting and erasing forces of trauma by creating an experimental, self-reflective form of (relative) connecting, is at odds with this new reading of Kincaid's work as interested in evoking a singular, opaque materiality that resists insertion in a (discursive) network of interconnections. But a more specifically Caribbean approach helps to understand why these two interpretations are not contradictory.

Literary theorist Patricia Donatien-Yssa (2007) suggests that a psychological or psychoanalytical reading along the lines suggested above does not do justice to Kincaid's writerly project (nor does postcolonial theory). She proposes to read Kincaid's oeuvre (notably *The Autobiography of My Mother* [1996]) in Caribbean terms, as an act of exorcism. She relates Kincaid's oeuvre to that of other Caribbean artists who confront "la blès"—"the suffering that is singular and foundational to Caribbean existence" (15, my translation). Donatien-Yssa describes it as a creole syndrome that manifests itself on a physical and a mental level. It emerged in Caribbean people who suffered enslavement and colonization, and is transmitted from one generation to another. It is therefore a communal, not an individual interiorization of centuries of violence. Caribbean artists use their art as a means of survival. As "la blès" cannot be undone, this art tends to "valorize the negative" (17, my translation). Kincaid's *Autobiography* can well be read within this frame, described by Donatien-Yssa as a mythic and mythological chronotope (18). Such a reading helps to understand her rage and aggression. These are all strategies to exteriorize "la blès," and to transform the effects of trauma into creative energy (18).

This useful intervention signifies a relevant warning not to rely too much on psychology or psychoanalysis, when it comes to the understanding of Caribbean writing.[18] Nor is a postcolonial approach (often informed by psychoanalysis) always helpful. Donatien-Yssa does not plead for an exclusively (mythological) Caribbean reading, though. She relates Kincaid's writing to a Deleuzian discourse by proposing to see her writing as a machine to produce desire (27), or a regenerative spiritual machine (29). Donatien-Yssa underlines the importance of spirituality as a foundational (and morally ambivalent) principle of Caribbean communities and individuals (28). Combining Deleuze and Caribbean epistemology, she states that Kincaid, by "projecting spirituality in the interstices of writing" (29), enters a process of the deterritorialization of spirituality, which brings her into what Deleuze and Guattari call "a world of pure intensities" (29). The intense spiritual sensuality of her writing (embodied by, for example, the awe-inspiring mother figure) would allow a confrontation with "la blès." Donatien-Yssa's study shows that, in a Caribbean context, a psychoanalytical approach that focuses on literary texts as individual constructions of identity may not be adequate, as Caribbean perceptions of self, health, trauma, reality, community, and language may strongly differ from what is assumed in Western psychology or psychoanalysis. However, Donatien-Yssa's own work also shows that Caribbean epistemologies are strongly syncretic, and can often be combined with European theories that (critically) respond to Western psychoanalysis, such as Deleuze's work. This creolization (which was already under way in work by Benitez-Rojo and Glissant) also means a transformative appropriation of these metropolitan theories.

Two insights that spring from this valuable argument are especially useful for my discussion of Kincaid's writing strategies. They support some of the insights I presented above, and help to see why Kincaid's response to trauma and her desire to return to materiality are intertwined. First, it is indeed illuminating to see Kincaid's writing as a response to trauma, and a desperate quest for transcendence—a transcendence that, in Donatien-Yssa's view, paradoxically, can only be reached by facing the horrors of "real" life.[19] Transcendence lies in facing the *truth* (her brother's dying, the truth of continuous harsh oppression and exclusion in the Caribbean), though that truth, as I explained above, might well *not* be related to a psychological, individual intensity of feeling. Instead of transcendence, we could also talk of liberation from oppressive relations and discourses, and a return to materiality. It is, moreover, certainly useful to see Kincaid's work as an effort at exorcism—that is, a manipulation of psychic energies, of affect, instead of the creation of an identity. Second, this reading helps to shift the focus from the (psychic) interior of the characters as the privileged site to obtain knowledge, to Kincaid's interest in the outside world, the body, the environment, as the spaces where truth manifests itself, disrespectful of all discursive efforts to contain it. Kincaid's narrator's inner turmoil is *externalized* in landscapes, colors, plants, that ultimately offer her a way to provisionally relate to the predicament of her brother, her family, and the Caribbean as a whole—even if this relation is no stronger than the shared color blue. Kincaid takes pains to point out that the relation between material object and painful event should not be defined as metaphor or analogy, or, to use Peircean terms, symbol or icon; often, the relation is indexical. I agree with Donatien-Yssa that a Deleuzian of new materialist approach would certainly be enlightening.

In the next chapter, which continues my debate of Kincaid's work, I will elaborate both points. For now, this seems to suffice as a provisional answer to the question on the aims of Kincaid's writing: it has the effect of working through pain, of exorcising trauma. It succeeds in doing so by testing the analogies between the human, the nonhuman, and matter, in a way that repeats and resists the logic of the Imaginary, to evoke a posthumanist stance. I can now also summarize Kincaid's relevance for ecocriticism. Her work helps to avoid an overly eager embrace of connectedness and empathy as the preferred literary devices. The final question remains whether this project also leads to the creation of a new sense of space, and, if so, *what* sense of space emerges.

LIVING SPACE

At the start of this chapter, I argued that Kincaid analyzes gardens as transnational landscapes, with very specific, transnational, colonial histories. This was especially clear in *My Garden (Book):*. At the very moment when Kincaid decides to use the garden as the perspective from where she can discuss colonialism, space, and identity, she turns away from what Glissant often sees as his enabling discourse: aesthetics, poetics, imagination. If Derek Walcott sighs "I have no nation now but the imagination," Kincaid doesn't agree. As for the imagination, Kincaid's position is: when I garden, my imagination fails me (and the discussion above will allow us to understand this resistance to the imagination and the Imaginary). "My garden has no serious intention, my garden has only series of doubts upon series of doubts" (6). Basically, the garden—and by extension, the world as a whole—is not just unknowable, but also unimaginable. Instead of a new imagination, Kincaid's *My Garden (Book):* offers a (seemingly) incoherent series of anecdotes, historical and botanical essays and personal musings, all mixed up, and, not unlike *My Brother*, all interspersed with sudden unrelated observations or exclamations that seem to contradict what has been said just before. I would argue that Kincaid emphatically downplays a perception of space as imagined space (in political geographer Edward Soja's terms), to privilege an exploration of what Soja (1996) calls *thirdspace*. To the definition of thirdspace, or lived space (as urban theorist Henri Lefebvre, Soja's predecessor, would call it) action is central. In Kincaid's work, the garden is very literally a practice, especially because Kincaid writes about garden-making rather than the resulting garden. Her accounts juxtapose reports about failed attempts at growing plants, or at creating a satisfying garden, with remarks about dying neighbors, leaking poison, and the loss of precious gifts. But the constant worrying *pleases* her: "The irritation to be found in the garden [. . .] will not lead to any loss of face; it will only lead to this question: What to do? and the happiness to be found in that!" (14). Clearly, Kincaid's garden is not about representation—not in the first place, at least. It is the site of a critique of (visual) representation, which is also a critique of the related desires to understand, know, and control.

Kincaid's critique of the imperialist image of the garden begins at the point where the landscape in its function as a technology of colonial representation naturalizes "its social and cultural constructedness" (Mitchell, 1994, 2). By showing how plants have been transplanted across continents, *My Garden (Book):* denaturalizes the newly shaped landscapes, and shows how these are the result of complex, layered, violent histories of ripping away and abduction; not just pleasurable places, but complex knots in transnational power relations. Thus, the text also refuses to present the landscape as a unity, in spite of the insistence of some theorists that the very concept of landscape implies unity; the landscape being understood as receiving its synthetic quality through the mediation of human subjectivity that wants to see it as a typical part of an undivided whole (Cosgrove, 1984, 13–14).

Kincaid understands the destructive colonial acts of Reason (naming, measuring, categorizing, controlling) as springing from an incomprehensible European pleasure. "[R]eason is a pleasure to them" (120). Instead of directly criticizing this desire, she states that her own pleasure is unpredictable and wily too. Her pleasure, however, is very definitely not related to the accumulation of knowledge, or to control and domination. Instead of the pleasures of knowing, Kincaid is sensitive to the pleasures of acting, which becomes very clear when she exclaims, in the quote above: "What to do?" Here, gardening is a desiring machine because it allows one to muse about possible future actions; it is an occasion for action, pragmatic planning, and performance.

This approach understands the power-infested field of postcoloniality primarily, in Deleuzian fashion, as a network of affects and actions.

This sensual, pragmatic approach to space goes together with a resistance to visuality, transparency, and therefore control too, that can also be found in Glissant. Two quotes may help to elaborate this resistance; both thematize shapelessness. The first quote is by the narrator in *My Garden (Book):*. She comments upon her garden's utter shapelessness, which puzzles her visitors:

> I had begun to dig up [. . .] parts of the lawn [. . .] into the most peculiar ungardenlike shapes. [. . .] These beds [. . .] were odd in shape. [. . .] I could see that they were odd [. . .] my beds were bigger, their shapes more strange, and they became so much more difficult to explain to other gardeners who had more [. . .] of an established aesthetic of a garden than I. (xiii)

Suddenly she understands, the "garden I was making [. . .] resembled a map of the Caribbean and the sea that surrounds it [. . .] the garden is for me an exercise in memory" (xiv).[20] This is not real or objective space, for it cannot be mapped. It is incomprehensible. Nor is it imagined or subjective space, for no existing discourse is capable of expressing it.[21] Kincaid cannot control, understand, or imagine it, she can only live it, build it, critically, as an anti-colonial anti-aesthetics. It is a form of *acting*—digging, planting, worrying—that is filled with shifting meanings.

Even if, in the wake of Hallward, I contrasted Kincaid's work to Edouard Glissant's later (more generalizing) work above, Glissant's earlier analysis of the condition of the world may help to situate Kincaid's approach. In the next quote, Glissant mourns the transformation of Caribbean landscape, in the wake of global capitalism:

> All these flowers have disappeared, or almost. [. . .] The land has lost its smells. Like almost everywhere in the world. [al] The flowers that grow today are cultivated for export. [. . .] But they are heavy also, full, lasting. [. . .] These flowers delight us. But they have no fragrance. They are nothing but shape and color. I am struck by the fate of flowers. The shapelessness yielding to the shapely. As if the land had rejected its "essence" to concentrate everything in appearance. (Glissant, 1989, 51–52)

Glissant deplores the loss of a shapeless world, and the advent of a world that is *merely* visible, *merely* perceivable, imaginable. Glissant's understanding of space, like Kincaid's, can be read as resisting the notion of imagined space. What is lost is not so much real space, but *lived* space: Glissant preferred the shapelessness because it demanded constant communal care (as he explains later on in the same text). This action is now lost, and space has lost its meaning—until, that is, the very moment when Glissant fills it again with meaning, with his analysis of the manner in which its materiality is shaped by its colonial past and transnational present. Glissant's point gains its full significance when it is related to Kincaid's statement about the shapelessness of her remembered Caribbean.

Clearly, the representation of the Caribbean as shapeless is more than an isolated, idiosyncratic complaint. It is a shared statement that is part of a project of a postcolonial redefinition. I would like to highlight the centrality of agency in this rewriting. The sense of space that is created in Kincaid's writing consists in an active engagement with a space that is experienced as transnational *lived* space, tying the Caribbean and the United States together. It is opaque, and irreducibly different. This sense of space is not predicated on an appropriating form of knowledge, but on another dynamics of desire: not the desire of territorialization, but that of the tensions between territorialization and deterritorialization (see Melas's contrast between *experiencing* and *knowing* difference above).[22]

Why consider Kincaid's literary *representation* of her garden lived space instead of imagined space? The reason is that this writing resists coherence and design. Kincaid's refusal to endow her writing with the kind of coherence and unity that allows for knowledge and control, and her explicit wish to disturb, confuse, and shock her readers, makes one understand her writing as a practice that is closely related to, and perhaps even part of her gardening. Instead of opposing writing (as an imaginative act) and physical acts of gardening, I would like to see these acts as related.[23] In all these practices, be they discursive or not, Kincaid's purpose is clearly not the creation of a unified image of the garden. All the practices related to gardening, even her writing, disturb unity, coherence, and the soothing certitude of imagined space. It is in this sense, to return to another question asked in the introduction, that "culture" and "nature" are related—if, by now, it still makes sense to use those terms. Writing (as a cultural practice) and gardening (as a cultural practice that relates to nature too) share the capacity to produce provisional, partial connections, *and* their disturbances; they are indexically related, and may intervene in each other. But neither explains the other. Both intervene in fields that cannot be known, only be acted upon.

Kincaid's focus on materiality and agency testify to her affinity to the posthumanist, (new) materialist theorizing. In the next chapter, I will continue my discussion of Kincaid's turn to materiality. Her search of tangible truth—preferably in human and nonhuman material existence—is not without its own problems and paradoxes.

NOTES

1. Kincaid's books about gardening and plants include the 2005 book *Among Flowers: A Walk in the Himalayas*. References to plants, gardening, and nature form an integral part of all Kincaid's writing.

2. Another critique of the notion of connectedness I would like to mention was presented by Hannes Bergthaler in a 2011 paper presented on the ASLE Conference in Bloomington, in which he discussed ecologist Barry Commoner's *The Closing Circle* (1971). Commoner's view of connectedness insists on the necessity of forms of closure (e.g., by membranes).

3. Christopher Herbert (2004) suggests that, in literary texts, coherence typically manifests itself in the form of "symbolic or metaphorical" relations (185). Kincaid's work shows that the principle of indexicality creates links without evoking a sense of coherence.

4. Remarkably, quite in contrast to Donnell, Lorna Down states that the text "replaces the arms' length treatment of an AIDS victim with the intimacy of the sister" (28).

5. See for example Lorna Down (2007) on the recurrent juxtaposition of "events" as a stylistic device in the text (22).

6. There is no end to the connotations of the color blue in the Caribbean. One of Olive Senior's poems is composed around the color blue, which refers here to the color of the banana leaves in the lyric subject's father's plantation, turned blue by the treatment against Leaf Spot Disease (2005, 86–87).

7. See Kincaid's (1988) essay *A Small Place*, and Hoving's (2001a) *Travelers*, 210–12.

8. See, for the link between the archaic and the modern, Shohat and Stam, 1994, 295, or see the large body of work on the modernist interest in primitivity—for example, North (1995), D. Miller (1991).

9. As outlined in her interview with Selwyn Cudjoe.

10. See Angeletta K. M. Gourdine (2003) for the notion of contamination, in her discussion of Michelle Cliff in a book on gender and sexuality in women's writing in the African diaspora.

11. I call this strategy productive, as Kincaid uses this same strategy in her more recent novel, that is, *Mr. Potter* (2002).

12. A comparable position is articulated in a 2008 volume edited by Craps and Buelens. In their introduction, they criticize the dominant Western psychoanalytical approach to trauma that emphasizes the need for empathy, which, according to them, is "a reaction that supposedly obviates any need for critical self-reflection regarding his or her own implication in ongoing practices of oppression and denial, let alone political mobilization against those practices" (Craps and Buelens, 2008, 5).

13. The critique of analogy occurs in different forms within cultural studies, and the study of race, gender, and sexuality. Miranda Joseph (2002) points out that "[a]nalogy works in opposition to the logic of abjection [which is] predominant in nation-building discourses" (as paraphrased by Cruz-Malavé and Manalansan IV, 5). Used in a political context, it helps to discern equivalences between social struggles, and therefore create political alliances between, say, those sexually marginalized and those racially oppressed. However, this political tool has important analytical drawbacks: it also *obscures* the specific connections between these affiliated parties and issues (Joseph 80). Kobena Mercer (2003) argues the same, with respect to race: "Although analogies facilitate cognitive connections [. . .] there is also the risk that they repress and flatten out the messy spaces in between" (191, see also Stepan, 1982). Because of this, analogy is a contested concept. "Analogy has also been a central feature of Fordist discourse, in the form of liberal pluralism of the nation-state" (Joseph, 80).

14. The semiotics of C. S. Peirce, for example, shows that analogy is a curious, mixed-up kind of argument. Instead of being a simple equation, it mingles the three basic types of argument (induction, deduction, and abduction, or hypothesis). The hypothetical moment in this argument might make one doubt its reliability. http://www.helsinki.fi/science/commens/terms/analogy.html, with quotes from, among others, Peirce's "A Theory of Probable Inference," CP 2.733–734, 1883 and "On the Natural Classification of Arguments," CP 2.513, 1867.

15. Miranda Joseph (2002) calls on Gayatri Chakravorty Spivak to explain that the workings of analogy depend on the problematic assumption of homogeneous, autonomous identities. With reference to the relation between the subject of consciousness and the subject of power, Spivak analyzes the effect of seeing these two as analogous: they are seen as "independent of each other" and as "complete unto themselves (each is posited as an internal continuity)" (80). The Wittgensteinian analysis of analogical interference in empathical introspection supports her analysis.

16. Jonathan Lamb even argues that slave narratives were modeled on the eighteenth-century subgenre of the "object tales," in which *things* relate their own story; for example, they follow the convention that the title mentions that this is the life of a certain object (a coin, a coat) narrated or written *by himself*, which is exactly the expression returning in the title of the narratives by Olaudah Equiano, Frederick Douglass, and Mary Prince (158).

17. Down and Covi both refer to Kincaid's writing in terms of concerning "being-in-the-world" (Covi, 19), or "'being' as having its own importance, as having an 'essence' separate from its construction by others" (Down, 2007, 26). Covi's approach, however, is predominantly informed by its focus on Kincaid's writing as the creation of a certain identity.

18. Mary Lou Emery offers a wonderful example of two opposing interpretations of a scene in *Annie John;* a merely psychoanalytical reading appears to be sadly inadequate (2007, 215).

19. Compare this with James Berger's (2005) observation that the imagination of modes of transcendence may be a response to the traumatic collapse of the Symbolic order (Introduction).

20. Susie O'Brien focuses on the aspect of memory, and reads the garden as a material record of the way places become inscribed by people and vice versa (2002, 171). I prefer to emphasize the aspect of the garden that escapes representation.

21. Pecic (2013) makes a comparable point on Glissant's and Kincaid's approach to space, but in a different theoretical context.

22. Though this differentiation is enlightening in the light of the above discussion, the two practices cannot be separated, as Peter Schwenger (2001) argues (see chapter 10).

23. Anne Collett's (2002) proposal to understand literary texts (in her case, too: Kincaid's texts) as friction (rubbing or making contact) is particularly felicitous.

Chapter Five

Exclusion

The Outside and the Outsider in Kincaid's Writing

Woven through Jamaica Kincaid's writing is the thread of her desire to touch base with solid truth. In his discussion of the way in which Caribbean writing aims to capture fragmentation instead of expressing organic wholeness, even Timothy Reiss forgets his own nuances when he is faced with Kincaid's desire for solidity. Reiss (2002) concludes his analysis of a fragment on flowers in Jamaica Kincaid's early work, after examining its irony and uncertainty as figures of loss, with a positive remark about the way Kincaid evokes solidity and wholeness: "they become the place of solid memory, enabling the writing that let the speaker 'feel myself solid and complete, my name filling up my mouth'" (1984, 82, qtd. in Reiss, 351). But is this search for a solid identity really what is at stake here? Reiss overlooks the fact that such rare moments, when the desire for discursive solidity and completeness are expressed, are always immediately contradicted in Kincaid's work.[1] Stability and solidity often seem characteristics to evade, not ones to pursue.[2] Reiss situates Kincaid's references to flowers within the conflict between the desire of a discursive, unified sense of self (an identity), and the difficulties inherent to obtaining it. In the context of Caribbean writing, however, the identitarian move toward a stable identity should not be mistaken for the only possible desirable, decolonizing emancipatory script. In Kincaid's work I see, rather, a tension between the desire to sarcastically adopt the discourses that identified and destroyed her, and the desire to move outside discourse altogether. In this sense, Kincaid's work is not emancipatory.

Let us catch up with the debates that have been raging around *My Brother* since it was published in 1997, to see how the tensions between the desire for solidity and the deconstructive narrative strategies of the text are theorized. Many critics, all fascinated by Kincaid's wayward work, have commented on the critique of binary forms of identity in her writing. The novel has been discussed, both very critically and admiringly, as an AIDS-narrative (often together with Patricia Powell's *A Small Gathering of Bones* and Lawrence Scott's *Aelred's Sin*), as an effort to work through grief, as another sequel in Kincaid's long autobiography, as another instance of her struggle with her mother, or as a pastoral romance of the narrator's family life in the United States (Paquet, 2002). Throughout, it has always been read as a text that tries to recreate a new sense of self through reconnecting with her Antiguan family. All critics comment on the ambivalence in this project of creating subjectivity, but most of them praise Kincaid's success in establishing a nonbinary, different speaking subject (Paquet)—that is, a pluralistic, postmodern, postcolonial (but not hybrid) self (Bernard, 2000). Or, as Covi (2003) proposes, a relational self, not articulated through identity politics but through a politics of intersubjectivities, which might be described as a prismatic subjectivity.

In this way, Kincaid is firmly associated with the project of moving away from the colonial notion of an autonomous, essentialist, disconnected identity, to embrace the notion of a multiple, relational self that is associated with postmodernism, postcolonialism, and more and more with Deleuzian and Glissantian materialism (as I have also argued above).[3] The postmodern and postcolonial *philosophical* notion of relational identity is then associated with the *psychological* and *moral* virtues of empathy and intimacy.[4] However, the relation between this philosophical debate (which often concerns the poststructuralist emphasis on difference—or *différance*—in the production of meaning) and a political or ethical practice is by no means clear. Relationality and/or connectedness are not necessarily "good." Most critics, fortunately, do not contend themselves with the mere celebration of Kincaid's creation of a relational self. Many move on to an analysis of her stylistic performance, which is then praised for its great poetic complexity. In spite of the wide acknowledgment of the ambivalences and paradoxes in Kincaid's work, most readings nevertheless aim at a certain closure. This closure is often sought by stating that Kincaid's style and production of subjectivity are happily congruous with the wayward, plural, harsh conditions of today's postcolonial, globalizing world.[5]

Instead of praising Kincaid's accurate understanding of postcolonial connectedness and identity, I will offer a more critical analysis of her writing and focus on the text's fascination with matter and truth. I argue that this fascination can be interpreted as a specific environmental view. For this reason, I am particularly interested in the role played by references to the natural environment in this text. At the same time, however, this environmental view is (inevitably) informed by notions of identity and sexuality that deserve closer scrutiny.

IDENTIFICATION: WHAT IS IT GOOD FOR?

In the last chapter, I suggested that Kincaid adopts two different writerly strategies to relating to her family and birthplace: creating distance and professing ignorance. A helpful essay by Derik Smith and Cliff Beumel (2006) helps situate the first strategy. In spite of the argument that Kincaid creates a plural self, this strategy works with a radical form of identity politics: a radical *separation* of "me" and "him," "us" and "them." Smith and Beumel build on Alison Donnell's caveat that the wish to celebrate postcolonial authors leads to a normative model of an "acceptable postcolonial response, which both chokes critics and arrests the possibilities for making meanings" (Donnell 1995, 101, qtd. in Smith and Beumel, 96). They argue that the "critic's duty is then, in part, to demonstrate the particularity of the author's voice by calling attention to its more problematic assertions" (97). The particularity of Kincaid's voice does not lie in its poetic quality, its complexity, relationality, or its truth. Instead, the two critics focus on the problematic aspects of Kincaid's voice by situating Kincaid's text in relation to the discourses that surround her, notably the discourse on homosexuality, with which she enters in a "somewhat dilettantish engagement" (97). With the help of an essay by Lee Edelman, they then interpret Kincaid's progressive, "fascinated 'unfolding' of Devon's life" (108)—an outing Devon himself has not wanted—as the heterosexual "drive to make legible (thus essential) the homosexual," which can be seen as "a function of ideological heterosexual identity construction" (109). In this vein, Kincaid can be said to "activate homophobic stereotypes in order to effect her self-construction" (110). This, then, is a kind of intimacy that can be really destructive for the person who was outed against his will, and as a result othered and excluded. Smith and Beumel's critique of ideology leads to a sobering response to the question how Kincaid's text effectuates a reconnection of the fragmented postcolonial: her texts would merely *enhance* the fragmentation by adopting exclusionary discourses and strategies.

This lucid analysis shows that literature is not merely concerned with the production of certain types of identity. It does more. It appropriates specific discourses, or intervenes in them. The overly eager morphological focus on identity (as stable, fragmented, plural) that one encounters in much critical work downplays this specific discursive historicity. At a price, since this neglect makes it harder to evaluate the text's cultural significance. One cannot judge the subversive or critical nature of a text if one doesn't identify the discursive networks it addresses. Kincaid's text does not just tie in with heterosexist discourses through its strategies of identification and othering. We also find a stereotypical link between homosexuality, death, and infertility that refers to heterosexist environmentalist discourses (Donnell, 1995;

Smith, and Beumel, 2006). Surely, as early as the 1980s, Jamaican writer Michelle Cliff wrote a moving text about this very strategy. In one of her autobiographical essays, she pondered how her passion for gardening was tied up with her mother's accusation that she (the lesbian, light-colored, ladylike daughter) was not able to procreate, to bring new forth life (Gourdine, 2003; Hoving, 2001, 251–53; see also chapters 2 and 6; see chapter 8 for an analysis that links fecundity to homosexuality). Is it possible to trace the link between these heterosexist discourses and environmentalism? To answer that question, I will consider the observations about gardening, gardening books, trees, and plants in *My Brother* that occur at many decisive moments in the narrative.

The difference with the more conventional literary use of botanical metaphors is striking. The narrator hears about her brother's predicament when she is reading a book on gardening she dislikes. She also reports reading the book again when she is visiting him (10, 11, 18, 19). The book makes her think about her brother's ability to grow things (19). She then reflects on the botanic metaphors she could use to describe his failure to be fertile and productive in any way, his rotting away, withering, and dying (19, 79, 113, 167). This explicit reflection, which questions the conventional use of metaphor, is an unusual literary device. Not all of Kincaid's literary references to nature are unusual. When the narrator pictures the relation between her mother and her children by narrating how the mother ruthlessly cuts down or burns down the trees that hinder her—even those planted by her son (12–13, 125–28)—her narrative fits in with conventional forms of nature symbolism. More complex, however, is the symbolism in the narrator's report that her brother died at the moment she was carrying two very cumbersome plants through airports, planes, and customs (177). The narrator adds, predictably, that these plants did not survive the transplantation. If the readers are sometimes invited to create a direct metaphorical link between the dying brother and the infertile plant, the tyrannical, smothering mother and the killing of trees, they are also invited to question and suspend these links. In this manner, the neat differentiation between homosexuality/infertility/death and heterosexuality/fertility/life is both evoked and destabilized. The readers are made aware that there is no inevitable, "natural" way to link the different members of her family, as none of the "natural" narratives of genealogy, maternal care, or fraternal/sisterly love are available. Connections can only be made haphazardly, provisionally. This critique of the natural family does not change the fact that Kincaid *does* adopt a heterosexist rhetorical strategy in her novel. At the same time, it should also alert us to a radically different discursive strategy Kincaid applies to address the issue.

One of the provisional connections Kincaid makes with her brother is established by the remark that the narrator can only really feel her sadness about her brother's death when she sees that the evergreens in her garden have been eaten by deer (102). It appears that, instead of allowing her to dismiss her brother from the realm of fertility and heterosexuality, the discourse of gardening offers the narrator a discourse to *relate* to her brother—to what she perceives as his failure in life, and his death. The deer have destroyed her own garden, indicating that she is affected too, though not by a nameless horror but by a specified herd of animals, deer. Thus, understanding comes from translating the distant event to an intimate, material experience. This affective strategy is in line with Kincaid's recurrent effort to come to terms with life by trying to move outside discourse altogether and touch base with the materiality of *her* being, a materiality she shares with her brother. The discourse of gardening expresses the basic material signification of her brother's existence and his death: his desire to grow things, his failure to do so, his rotting away and dying. The narrator needs basic, nonpsychological, nonhumanist words to address the materiality of this reality, in which she also grounds her own life.

Her interest in decay and corruption, evoked by her brother's death, does not just serve her homophobic desire to radically *other* him (though it does do that too). It is also a move toward a posthumanist perspective, and in its acceptance of loss it is close to dark ecology (though it does not share dark ecology's queerness). At the same time, this acknowledgment of rot and decay as an inevitable part of life is kindred to the "aesthetics of the earth" outlined by Glissant:

> An aesthetics of the earth, for Glissant, does not stem from a simple appreciation of beauty, especially since so much of the environment of composite peoples has long since been ravaged by colonial violence. But it is precisely the seeming inappropriateness of aesthetics in the context of waste and rupture that can enable a regenerative response. (DeLoughrey et al., 2005, 22–23)

This quote is especially relevant as an astute example of an inappropriate aesthetics in Caribbean writing, often by women, that brings together sexual, political, and environmental discourses on corruption, infertility, the refusal to procreate, rotting, and degradation. Kincaid's work represents one eloquent example, but it is certainly not the only one available. For example, in Dutch-language Surinamese (migrant) women's writing, these forms of degradation are analyzed as a postcolonial condition that characterizes society as a whole, rather than the characteristics of the abject (see chapter 6, "Dwelling"). Ironically, it is the very evocation of a living, dying, and corrupt nature that enables Kincaid to bypass the strategies of empathic identification from the *inside*, and adopt a posthumanist perspective from the *outside*. We are reminded of Donatien-Yssa's (2007) statement that Caribbean art that confronts "la blès" "valorize(s) the negative" (17).

In chapter 4 I argued that, instead of empathy and identification, Kincaid uses two other strategies of relating to her family. Firstly, by taking her distance, and secondly, by claiming ignorance. It is only after having successfully taken her distance by clarifying her identity through a binary opposition to her brother's homosexuality that the narrator can adopt the second strategy (claiming ignorance). She does this by continuing her efforts to clarify and *reduce* her own identity and that of her brother by imagining both, on the level of plants and animals, as mortal organisms. At this physical level of experience, opened up by her explicitly imperfect and provisional work with analogies and counteranalogies, she is able to reflect on the inexpressible, sad connectedness of her brother's and her own embodied existence. In other words, Kincaid's text needs the identitarian, homophobic moment of outing her brother, in order to bridge the gap between her project of taking distance from her family, and her project of moving outside the realm of discourse and knowledge altogether. Only through this series of stark reductions, and by finally evoking the most basic, material level, is she able to realize a sense of connectedness.

I can situate this analysis within a broader context by going back to the debate on empathy and sympathy in the preceding chapter. Like Max Scheler, literary scholar Jonathan Lamb also reveals a connection between sympathy and death. Lamb's evaluation of the concept of sympathy is particularly useful for ecocriticism, and helps us understand Kincaid's reimagination of herself and her brother as mortal organisms. In a magnificent essay, Lamb (2001) traces the history of the notion of sympathy between humans and the nonhuman from Pythagoras (who related this sympathy to his belief in the soul's transmigration to other forms of existence, after death) to non-Pythagorean British philosophers such as Locke and Hume. In spite of their refusal of the Pythagorean idea of the transmigration of the soul, these philosophers do not dismiss the idea of sympathy. Lamb describes Hume as "the first British philosopher to investigate the social and psychological subtleties of sympathy" (135).[6] Descartes is placed in opposition with his British colleagues' interest in the embodied self, as he holds a strongly dualist view of man's body and soul, and "was also notorious for his unkind attitude to animals, whom he counted nothing but machines" (145). For Hume, in contrast, the possibility of sympathy between men and animals emerges from a shared fear of death. Thus, Lamb shows that sympathy with other organisms and substances was predicated on the "the dissolution of the body" (135), as this is what metamorphosis (as explored by Ovid and others) comes down to. This process was, as Ovid's narratives and those of his followers show, far from smooth or happy, and often motivated by terror or hatred. The association between sympathy, metamorphosis, and death, returns in present-day novels such as Coetzee's *Disgrace*, where the likeness with animals (in this case, dogs) results from complete powerlessness and humiliation (138).

In his conclusion, Lamb says that "the enlargement of sympathy (to the nonhuman world, IH) discloses relations that are neither comfortable nor sociable" (166). But comfort, of course, is not what Kincaid is after. Her strategy of connecting through identification with nonhuman mortal organisms is highly uncomfortable, but it is successful.

Is it possible to follow Smith and Beumel's illuminating example and situate Kincaid's references to nature as matter within a broader discursive context? In this narrative, obviously, nature is not the dense network of colonial discourses and strategies analyzed in *My Garden (Book):*. Nor is it the sublime alterity evoked in *Cereus Blooms at Night* (chapter 2). It is also not the source of a wealth of erotic and spiritual discourses (chapter 3). The stark reduction of nature could be related to a problematic reductionism in biology and ecology, which, according to environmental theorist Alan Marshall, can be found in both systems theory (Eugene Odum) and holistic theories (Lovelock's Gaia theory). This reductionism "reduces the actions of populations, communities, and ecosystems first to populations and organismal genetics, and from there to molecular biology and organic chemistry" (Marshall, 2002, 248). Ultimately, then, this approach to nature makes it impossible to *know* nature in its full complexity. As such, it seems to be an effort to evade knowledge about life, rather than an effort to understand it.

This impossibility of knowing need not be a problem for literature, which is often concerned with the tensions between knowing and not-knowing (Caruth, 1996), and often more concerned with the manipulation of affect than with the production of truth. If Kincaid's text pushes its reductionism until it hits what appears to be a most basic truth of life and death, we may well argue that this is no truth at all, but merely another construction within certain, problematized ecological discourses. Yet Kincaid's novels need these fictions, these reductions, to arrive at a seemingly nondiscursive space where they can affectively relate to the exteriorities of the fragments of a painfully divided life.

HIGH ON NATURE

To what extent can we understand this radical, reductive exteriorization as the reenactment of a traumatized condition, instead of a regenerative response?

A famous essay by Homi Bhabha, who is interested in the schizophrenic condition of the postmodern, suggests we should see it as reenactment. Bhabha quotes Fredric Jameson's account of the

> breakdown of temporality [that] suddenly releases this present [. . .] engulf(ing) the subject with undescribable vividness, a materiality of perception properly overwhelming. . . . This present of the world or material signifier

> comes before the subject with heightened intensity, bearing a mysterious charge or affect . . . which one could just as well imagine in the positive terms of euphoria, a high, an intoxicator. (Jameson, qtd. in Bhabha, 2004, 215)

Bhabha reads this as the "anxiety of enjoining the global and the local" (216). It is the same condition that is explored in Kincaid's work, as the pathology of formerly colonized people, who are victims of the "breakdown of temporality." In Kincaid's writing, too, it is impossible to differentiate the *negative* dimension of this condition from the *positive.* Instances of this are sometimes quite innocent, as when Kincaid's narrator explains that she loves a certain tree because of "so personal an experience as when my mother would not allow me to do something I particularly wanted to do and in my misery I noticed that the frangipani tree was in bloom" (1999, 169, also 1998, xiii). The pathological dimension in this experience becomes even clearer in the account of an intensified experience of her surroundings that occurred during a physical punishment (1996, 8–11). Indeed, in moments of distress, temporality may be halted. It is true that gardening can be an "exercise in memory" (1999, xiv), "Memory is a gardener's real palette" (1998, xvi), but Kincaid is also very much interested in the (often painful) moments when plants are suddenly disengaged from history and take on an intensity that seems transcendent and life-saving.

Consider the following moment, when the narrator reports how, to her great joy and confusion, she encounters in the glasshouses in Kew Gardens:

> The most beautiful hollyhock I had ever seen. [. . .] It was a most beautiful yellow, a clear yellow, as if it, the color yellow, were just born, delicate, at the very beginning of its history as "yellow," but when I looked at the label on which its identification was written my whole being was sent a-whir. It was not a hollyhock at all but *Gossypium*, and its common name is cotton. Cotton all by itself exists in perfection, with malice toward none; in the sharp, swift, even brutal dismissive words of the botanist Oakes Ames, it is reduced to an economic annual, but the tormented, malevolent role it has played in my ancestral history is not forgotten by me. (1999, 110–11)

There is great tension between two different views. On the one hand, the flower appears as the junction of the histories of colonialism, economic exploitation, destruction, and murder. As such, this is a good example of a representation along the lines of the discourses of interelatedness. These memories, histories, and discourses, triggered by the name *Gossypium*, are presented as real, immediate, and as actual as the flower before the narrator's eyes ("This is not a fiction; this all occurred to me while seeing this particular flower in bloom in Kew Gardens" [112]). On the other hand, the flower's isolated presence seems to exist outside these discourses. Its yellow color is not part of history, but it has its own history which is starting only now.

Moreover, the flower's beauty is experienced as transcendent in intensity. As a *thing*, in Bill Brown's sense,[7] it has its own curious temporality—it occurs *before* it is recognized as the *object* that is overdetermined by the sheer weight of history, but it is at the same time an excess—"a sensuous presence or [. . .] a metaphysical presence" (5), and as such, it also comes *after* the object. This afterlife is exactly what motivates this observation. It captures the lasting emotional shock caused by the sudden transition between the supposed hollyhock's (or perhaps rather its yellowness) *thingness* and its role as an object. The shocking ambiguity cannot be resolved, and there is no way to escape the painful simultaneity of the two intense emotional responses. After differentiating the thing's temporality as both "before" and "after" the subject, Brown highlights this irresolution by arguing that matters are even more complex: "this temporality obscures the all-at-onceness, the simultaneity, of the object/thing dialectic" (5). The two visions on the cotton flower (as "real"/thing, and as historical/object) are not reconciled, but are shown as so radically opposed that their confrontation results in physical shock.

One might recognize Kincaid's desire to free a space outside of colonial and imperialist discourses, while acknowledging the powerful and very real effects of those histories and discourses.

> What to call the thing that happened to me and all who look like me?
>
> Should I call it history? If so, what should history mean to someone like me?
>
> Should it be an idea, should it be an open wound? [. . .] Is it a collection of facts, all true and precise details, and if so, when I come across these true and precise details, what should I do, how should I feel, where should I place myself? (114)

By questioning the usual definitions of history, the narrator tries to find the meaning of history for those who are descendants of slaves. She also ponders (and does not dismiss) a concept of history as a noncausal collection of separate, specific facts. Here, she proposes the same view as when she watched the yellow cotton flower without knowing its name: an effort to grasp the event or thing in isolation, in its immediacy, in its materiality.

If we adopt a psychological perspective, we could see Kincaid's dismissal of causality as mirroring the schizophrenic inability to recognize causal relations, and the inability of people who are deeply traumatized to relate to the external world. Should we then read Kincaid's representation of isolated fragments and facts not as philosophical reflection, but as an effort to linger in the space of colonial trauma? Are Kincaid's analyses of flowers as junctions of global processes the moments where she heals that condition, by embracing the discourse of relationality? One answer might be that Kincaid's focus on isolated things and events is indeed mirroring the schizophrenia of the postcolonial condition (Hoving, 2001a, 211). Her sudden "high," caused by the cotton flower that briefly transported her beyond history, repeats, "quotes," the schizophrenia of colonial trauma. Kincaid's strategy of inhabiting the discourses she criticizes by quoting them is a very visible aspect of her writing (Hoving, 2001a, 210–12). Instead of adopting the supposedly healing strategy of reconnection, Kincaid prefers to explore the spaces of the traumatized. Donatien-Yssa helps us understand this as a healing strategy in its own right.

This turn to materiality can be understood as the productive alternative to what Peter Hallward criticizes as a singular poetics. Writers who offer a general, singular poetics in which the specificity of their condition within (post)coloniality or globalization threatens to get lost, such as the later Glissant, are said to occupy one position within the field of Caribbean writing. This tendency is counterbalanced by those writers who occupy a seemingly opposed position, such as Kincaid, and try to restore the materiality of the fragments and shards that have lost their meaning in the violence of globalization. There are two ways to restore materiality. One way is to reinsert an element or event in its specific political and economic context. The other way is to affirm its isolated, individual, material quality. The latter strategy, seeking to place the thing outside discourse, can well be seen as a necessary countermove to the flight into the imagination of totality.

Nevertheless, a broader understanding of the concept of singularity would lead to the question whether Kincaid's evocation of materiality is not also marked by singularity (as Melas's analysis, discussed in chapter 4, would suggest). Timothy Clark (2005), in his study on the concept of singularity in Heidegger, Gadamer, Blanchot, and Derrida, adopts a different approach than Hallward. Clark refers to the very discontinuity and isolation that also characterize Kincaid's description of "things." He understands singularity as "the implicitly anti-determinist, acausal force of eventhood [. . .] [which] may involve a kind of acausal discontinuity" (3). Especially useful in this context is Clark's reference to Attridge and Readings, who relate singularity to non-rationality and a resistance to grand narratives, clarifying the way in which singularity opposes reductive forms of identifications that might be the undesired effect of a culturalist academic identity politics. The insistence on corporeality and situatedness is a key point in the research of singularity. In her introduction to the edited volume *Singularity and the Poetics of Translation*, Birgit Kaiser observes in all theoretical approaches to singularity a shared "concern for non-abstractable, corporeal, situated, pluralizing differences" (5). Or, as Derek Attridge has it in the same volume, in contemporary thought "singularities are sites of resistance to the universal" (6).

In these studies, singularity is addressed as a critical force that counters reductive discursive fixity. I would, however, hesitate to situate Kincaid's writing strategies within this framework. The reason is that Kaiser (2014) takes care to point out that singularity "cannot be reduced to sheer materiality—as inaccessible or opaque" (6). This is nevertheless precisely what Kincaid's work *does* offer.

Attridge (2014) might recognize in Kincaid's writing Paul de Man's notion of radical singularity, "so absolute that it is unknowable" (62). Remarkably, Attridge rather points at the affinity between De Man's definition and Hallward's critical description of what he sees as singularity in postcolonial theory (such as Glissant's concept of Relation). Kincaid's evocation of isolated chunks of matter seems to be radically opposed to that generalizing, postcolonial form of singularity.

Scholars such as Kaiser and Attridge insist that the most productive understandings of the term singularity would acknowledge the fact that singularity is not completely unknowable and unrelated. However, in Kincaid's writing, the sudden shock with which she calls forth sheer unknowable matter only materializes because it evokes the radically unknowable other. It is in this uncompromised way that Kincaid's evocation of the materiality, thickness, and unreadability of the cotton flower function as the moment of resistance against appropriating discourses.

It is true, however, that Kincaid's writings do not restrict themselves to this particular strategy, and that they apply literary devices that create analogies too. Her texts offer a plurality of quickly alternating strategies of disconnecting, distancing, reduction, inserting into history, and dismissing history, and none of these strategies gains the upper hand for long. Thus, Kincaid's texts resist both the pull of discursive appropriating and the pull of a singular, universalizing poetics criticized by Hallward, as well as the collapse into the shock of materiality. Adopting the "high" of facing nature as frozen atemporality is as much a part of the process of working through trauma as the acknowledgment of the historicity of nature. So is the reduction of other people's identity. It helps to deal with one's own instability, even if that means the adoption of exclusionary discourses that work against the interests of others. As such, her texts should not be read as an enlightening philosophy on the condition of the world (though they do intervene in philosophical debates), or as an exemplary imagination of a relational self that is healing its colonial trauma, but in the first place as the textual manipulation of affect. I will elaborate on this in the next section.

WRITING AND THE CRITIQUE OF IDEOLOGY

Some critics offered an evaluation of Kincaid's literary strategies by expressing a harsh criticism of the text's homophobia. Remarkably, Smith and Beumel, who pointed out her heterosexist strategies with the utmost clarity, refrained from doing so. Testifying to a nuanced understanding of not just the social, but also the affective dimensions of writing, they avoid accusing Kincaid of heterosexist essentialism (or, for that matter, of the imperialist tendencies suggested by her "righteous foreign aid" (111) narrative). Instead, they explore the signification of this "unleashing of the dangerously ideological energies late in the text," by returning to the text as a whole. On the one hand, as Edelman explains, homosexuality serves as the means to stabilize heterosexual identities. On the other hand homosexuality also has "the power to signify the instability of the signifying function *per se,* the arbitrary and tenuous nature of the relation between any signifier and signified" (Edelman, 6, qtd. in Smith and Beumel, 110). At this point, Smith and Beumel join the other critics who point out the contradictory nature of Kincaid's project of articulating a self that at the same time dissolves itself.[8] They read this contradiction as postcolonial, while emphasizing the colonial impulses in this process. I prefer to see the most remarkable effect of this project neither in the destabilizing of her identity, nor in the presumed success of pluralizing it, but rather in the reduction of identity to its most basic, organic form. Even so, I admire the sensitivity to contradictions Smith and Beumel put forward in their analysis, and the courage it takes to define these contradictions. Their acknowledgment of this complexity leaves room for a nuanced consideration of the intricacies of the writing process itself, instead of just a critique of ideology.

With the reading presented by Smith and Beumel, we have left the psychoanalytical reading of Kincaid's text behind us and adopt a deconstructive approach. In a discussion of Charles Johnson's work, Peter Hallward (2001) suggests that "Johnson's fiction does not much convey a knowledge of things as rehearse (or as Deleuze would say, repeat) the process required for an escape from knowledge *of*" (170). In Johnson's case, this escape from knowledge is effectuated by constant deferral. Kincaid's text uses the same strategy, but also accomplishes this effect by explicitly questioning signification, and through constant references to nature as discourse's outside. Hallward's remark sheds a welcome light on the other functions of literature, besides the creation of identity and the intervention in social discourses. Literature also dissolves and deconstructs, without necessarily relating this deconstruction to a constructive project. Literature also contains forces that oppose signification and refuse representation.

It makes sense to differentiate, in a text, between the energies that construct and those that deconstruct. For the writer, writing is a highly ambivalent process that does not necessarily reconcile the intensely opposed forces of creation and destruction. Proceeding from this insight, it makes little sense to focus on one isolated moment in a text in order to accuse the whole text of homophobia. My suggestion would be to approach the act of writing (from the perspective of the writer) as a basically *amoral* manipulation of narrative energies. A manipulation that makes use of images, patterns, and strategies that are *morally*, *politically*, and *culturally* significant.[9] Theorist can—and should—situate all these elements, aspects, and discourses—as Hallward and Smith and Beumel do. It is not very helpful, however, to offer a moral or political evaluation of any *single* part of a text. It would appear that writers use politically charged means to supplement their imperfect chains of connections, for good or for bad. Instead of taking a writer to task for needing an ideologically questionable connective or supplementing device, we might first want to analyze the *reasons* why that specific device, in this social and cultural context, would have such connective currency. In this case, we can sum up the answer to that question as follows. Homosexuality serves as a point of reference for the stabilization of heterosexual identity, and therefore as a means to create distance. Homosexuality has the power to signify the instability of the signifier (Edelman), and homosexuality evokes and questions a whole discourse on nature and the natural. This allows the author to address the most basic questions about nature, life, and death in new ways. This last point, already in part discussed above, will be taken up again below, in relation to another occurrence of homosexuality in Kincaid's writing.

My reading implies that every text contains constructing and connecting, as well as deconstructive and disconnecting movements. As such it opposes readings that focus on only one dimension of a text (those that accuse Kincaid of homophobia, and those that praise her for her creation of a plural, relational self). As Chris Bongie (1998) mentions, in relation to Glissant's notion of relation, we should read *all* elements of the combination *identité-relation*, including the hyphen. Identity and relationality exist in relation to each other, and we should expect to find the workings of both in a text. In a comparable spirit, Jean-François Lyotard makes clear (in contrast to Deleuze) that he does not just want to celebrate the "excellence of the movement of desire as against its fixation" (Bennington, 1998, 34). The process of creating structures, organizing and unifying is also shaped by libidinal economy, just like the processes of deconstructing, dissolving, disconnecting. In his *Libidinal Economy*, Lyotard provocatively claims that we do not need "a new morality with this consolation [. . .] that the libidinal band is *good*, that the circulation of affect is *joyful* [. . .] it isn't an ethics, that or another, that we need" (Lyotard, qtd. in Bennington, 34). Within the area of postcolonial theory, one might not quite agree with this view on ethics, but I do subscribe to the suggestion that, when it comes to understanding what writing is about, we do not need an ethical approach *too soon,* nor do we need a restricted understanding of what is "good" in a text.

Kincaid's writing is first and foremost a *writerly* project, one that sometimes tries to dissolve itself by exploding language's infinite signifying forces, in the face of the materiality of nature.

NATURE AND SEXUALITY

One of the paradoxical ways in which Kincaid finds a space for reconnection (i.e., reconnection with her family and through her family to the Caribbean), is her approach of nature as bare materiality, evading history and morality. But this is not the only role nature plays in her writing. I will conclude my discussion of the place of nature in Kincaid's work by turning to a curious reference to homosexuality in a story in *My Garden (Book):* that is dedicated to the narrator's as yet unsuccessful efforts to create a real family. These efforts are immediately bound up with her frustrated (dis)connections with her mother, brothers, and father.

A naive reader might assume that references to nature "naturally" provide a wealth of discourses of connectedness, even for those wary of politics. For nature seems to connect and copy in glorious indifference to all discourses of connectedness. It is in its *nature*, as Walter Benjamin suggests in a 1933 essay: "Nature creates similarities. One need only think of mimicry" (qtd. in Taussig, 1993, 19). However, it is not at all apparent what it *means* to say that nature is connecting and copying, even in relation to families. Organicist defenders of the natural (a group to which Benjamin obviously does not belong) understand the truism that nature is an interconnected network in a completely different sense than, say, posthumanist queer theorists. If the former like to see nature's wholeness in terms of harmonious heterosexual bonding, queer theorists find in nature's strategies of connecting and mimicking a liberating critique of heterosexist identitarianism. Following Halberstam and Livingston (1995), we can say that for the humanist, the human family is the preferred trope through which human beings are imagined as connected. For Kincaid, however, this trope does not work, because for her the family is potentially a space for cannibalism. Halberstam and Livingston have different reasons to problematize the trope. It implies, they explain, a notion of heterosexual, reproductive sexuality that is seen to embrace the whole natural world, since it imagines the world as tied together by genealogical relations that can be traced back to the tiniest unicellar organisms as common ancestors. For the posthumanist, however, heterosexuality is not the decisive factor to connect all organic life: "the ecology of interdependence problematizes the role of fucking in the life of species" (11). This attractively irreverent statement is corroborated by a variety of arguments. I would like to summarize these arguments freely as follows: developments in global ecology are not only caused by technology, but also by organic events (e.g., farting cows) (11), so that the connected world should be seen as a network of the organic *and* the anorganic, the "natural" *and* the "cultural." Within this global ecology, human bodies can be theorized as a "mosaic of microbes in various states of symbiosis," as Dorion Sagan suggests (qtd. on 11)[10, 11]—as multiplicities instead of unified systems, that can therefore not be mapped onto one specific slot in the "Great Chain of Being." For their existence and reproduction, bodies depend on material and discursive networks, including capitalism, institutions, sexuality. This "means that the umbilical cords that supply us (without which we would die) are always multiple" (17). The posthumanist connected world exists as multiple material and discursive networks, in which everybody is an open-ended multiplicity in itself, material *and* discursive; human *and* plant *and* animal; sexual *and* asexual.

This posthumanist analysis is based on a deconstruction of traditional notions of human sexuality, which is one of the reasons why I evoke it here. The comparison between organicist and queer approaches to interconnectedness shows that any play with analogy, connecting, and copying ties in with a specific discourse on sexuality, economy, and politics, also when it is inspired by nature. Even if she is no queer theorist, Jamaica Kincaid is very much aware of this connection. In spite of that, she needs to address the notion of the family, to effectuate an ambivalent form of reconnection.

Above, I suggested a provisional answer to the leading questions of this chapter—Do Kincaid's literary strategies in *My Brother* help restore a sense of place? Do they bring together some of the aspects that were torn apart by colonization and its aftermath, so that it becomes possible to relate to the degraded Caribbean (again)—perhaps through reconnecting with her family? I argued that the narrative does indeed reconnect—but only by redefining human and nonhuman life in reductive biological and physiological terms. I will now consider other strategies of reconnecting, besides reduction, suggested in Kincaid's work. While I have shown above how Kincaid used references to *nature* to relate to homosexuality and death, I will now show how she uses references to *homosexuality* as the means to fill up the gaps that makes it impossible to create a (heterosexual, "natural") family. In other words, her desire to both problematize and restore the family implies a problematizing of sexuality and nature. This curious moment occurs in *My Garden (Book):*. The moment is significant, as it links homosexuality to home-making and gardening, while reflecting on the dichotomies of in-/outside, inclusion and exclusion.

The chapter in which the unexpected reference to homosexuality occurs is called "The house." This chapter relates how the narrator acquired the Vermont house where she now lives with her family, and explains how hard it is to turn it into a home. It is a highly associative account, in that unmistakable Kincaidian style, in which memories occur in turn with short detailed descriptions. These are alternated with evaluations and reflections about possible causes or meanings, which, more often than not, remain unresolved. Most striking in this chapter is the play with the notions of inside and outside. These concepts are linked to the United States and the Caribbean, with the narrator currently living in the former and having grown up in the latter. "But I do not believe that I know how to live in a house. I grew up outside"

(28). "My house looks quite like the outside in which I grew up" (29). "The inside of my house looks like my yard; it is smudged with dirt, it is disorderly for an inside of a house, though it would look wonderful and memorable if it were the outside of the house I grew up in" (31).

In general, the home tends to be projected as the opposite to the destructive, homogenizing forces of globalization.[12] In response, Miranda Joseph (2002) argues that a romantic appraisal of the local (the community, the nation, the home) as the preserver of diversity, as opposed to the totalizing and homogenizing onslaught of globalization, neglects the fact that localization is also a process of capitalist evolution (74). The social formations of the home, too, are produced by one's participation in the (both global and local) structures of production and consumption. The notion of the home is complicated not just by class, but also by race and gender.[13] Kincaid's analysis of her postcolonial migrant's home can be understood to express this same insight: the home cannot function as a closed, authentic "inside."

Following Joseph's lead, I choose to read Kincaid's account of a continuous, half-failing effort at buying property and home-making as an exploration of the forces that have to be mobilized to create a postcolonial home. To begin with, the narrator is obviously not in control of these forces. While the house is in effect sold to the narrator and her husband, the former house owner's heirs never really succeed in letting go of the family property. The new house owners have to promise the heirs that if they ever were to sell their house again, they would offer to sell it back to them first (22). The children ask them for lost objects (22–23), and want them to take care of the piano until they have found a place to accommodate it (26). In addition, the narrator is often reminded of the former owner by an object he left, or an inexplicable smell of smoke that sometimes appears to linger in the room where he died, and which is now the narrator's study.

The failure of this project of possession is not only the fault of the former house owners' family. The narrator herself was not unequivocally aiming for possession, to obtain capital, in the first place. In other texts we find a confirmation of Kincaid's narrators' deep-seated mistrust of the wish to accumulate capital, which originates in their awareness of capitalist colonial history and its treatment of human beings as capital, like bales of sugar.[14] The narrator's wish to live in the house might be read under the nominator of desire (of being in a certain space), rather than under that of identity (of being a house-owner, of being secure, settled, a full citizen). The narrator inhabits the house in the way she inhabits dominant discourses. That is to say, she is aware that they do not define her, that she is only quoting and assuming them, provisionally inhabiting them. She insists on expressing the differences between herself, the house, and its history. Her unease with capitalism keeps her from fully identifying with the home.[15]

Her position is not only different in an economical sense, but also in terms of genealogy. Whereas the possessive children of the house owner cannot really let go of their memories, the narrator remarks, "I cannot believe that my children will return to this house shortly after I am dead [. . .] and ask the new owners to retrieve [lost recipes]" (22). This remark links the theme of owning property to the theme of genealogy. There is a second way to read the narrative of the home, then. Not as a capitalist account of creating family property (the family being the basic component of consumerist society), but as a genealogical account. The home is then read as the site of reproduction and procreation, which is defined by the specific relation between parents and children. Clearly, the narrator's children feel different about their parents than the children of the former owner of the house. A few lines below, we read that the narrator is also unable to imagine that her children want to admit that they are the product of their parents having sexual intercourse (23). Even the ties between parents and children are ambivalent and uncertain.

The chapter offers two failed narratives, then. One is the story of the unsuccessful effort to fully possess the house, the other relates the failure to create a stable family order in which the genealogical aspect is secure. I propose to read the seemingly inconsistent reference to homosexuality—which I will describe below—as a supplement to repair these failed narratives.[16] At a certain moment, right after the narrator (at this point a mother of two) exclaims that she does not know what to teach a child (a home also being the site of pedagogy), she finds herself with her husband in the kitchen. Though their four-year-old daughter is actually with her in the kitchen, the mother's parenthood does not seem to have fully taken hold yet. "There is a moment when you are the parent of a child and you [. . .] still think you are alone in the world" (32). The child is literally out of place. Apart from being unseen, she is sitting on the kitchen table, which she is not allowed to do. Nevertheless, as Smyth and Croft (2006) state, Bachelard suggests that there is an intricate link between the perception of the home and the child: "the preoccupations of childhood and adolescence determine the role domestic space plays in the imagination of the mature subject" (24). Without the child, there can be no home, no family. Without thinking of the child, the parents are discussing a man's homophobia. Suddenly, unexpectedly, the child enters their conversation, demanding to know what a homophobe is. The child makes herself known as a speaking subject, and as a subject of parental education. By doing so, she re-creates in this moment the people in that kitchen as a family.

The moment is triggered by a discussion of what is, in Miranda Joseph's (2002) words, a binary narrative of exclusion. In response, the girl's father offers a narrative of inclusion that, at first sight, seems to be based on analogy. The narrator gives the father the floor to let him explain what homosexuality is ("Two men meet and they fall in love with each other and they decide to get married" [32]), and what homophobia ("The people who are afraid of two men who decide to get married, they are homophobic" [32]). He explains homosexuality to his daughter in terms of traditionally heterosexual family institutions (they "get married"), and explains homophobia by situating it in the category of "phobias," "fears." The father does not add any moral judgment. The narrative focuses on institutional aspects, implying that these institutions are accessible to everyone, rather than on an inclusive ethics of identification or analogy. The girl responds in kind, immediately grasping the essence of the father's story; that family structures are produced by rules and laws. "'Just tell me one thing, who wears the veil?' Her father said, 'They decide.' And Annie said, 'Oh.'"

Annie's response—"Oh"—reinforces the absence of moral evaluation in this account. In this short exchange, the word "decide" is used four times. What is emphasized by this repetition, is that the rules that create couples and families do not work naturally or automatically. They are *called upon*. In this account, people are able to decide whether and how one wants the rules to apply. The freedom to decide and to negotiate is central in this text, which presents family structures as performance (or even transvestism), rather than naturalizing them as community or home. At the same time, the contested nature of this performance is mentioned.

In spite of this textual production of a family, the family is still described as lacking. I suggest that the family does not yet function because the narrator, the mother, cannot acknowledge her genealogical ties—and understandably so, in the face of her past. She still thinks that she is "alone in this world" (32), which is of course exactly what she had to be to survive her mother's destructive appropriation of her children, the maternal cannibalism discussed in chapter 4. In contrast to the former house owners, she lacks the security of a representable, relatable genealogy—to such an extent that the writer's oeuvre can be read as an exercise in re-creating the impossible ties between parents and children that are incapable of acting as such, *without* taking recourse to the nineteenth-century discourse of genealogy. It is not surprising that the narrator lets her husband respond to her daughter. She may be as anxious as her daughter is, to learn how her husband will fill the gap where the (grand)mother should have been.[17]

This dialogue creates a family. A father takes on his role as educator, and, in doing so, acknowledges his daughter as a speaking subject in her own right. No stable inside is produced. The home is not created by inclusive gestures. The creation of family ties is a matter of open but contested negotiations. Father and daughter do not bond through empathy or analogy either, but through dialogue. The paragraph that follows, closing the chapter, shows that the result will be necessarily unstable. It opens by turning back to the question asked at the beginning of the exchange about homophobia and homosexuality: what to do in a home, what to teach a child. A few lines below the desire for a blueprint is repeated, "how we wish that someone, but perhaps Robert Woodworth [the former house owner] in particular, had given us a recipe for how to make a house a home" (33). Woodworth's success shows in the success of his children, who are living "a life of stability" (33). At the same time, however, the readers are made to understand that the narrator takes her distance from this success, as "all three of them are people we would have been afraid of becoming, if only because the stability of such lives remind us of our own youth" (33). Stability, order, nonambivalent belonging and unambiguous genealogical chains, tied to the parental home, are opposed to the characteristics of the yard. There, we find incessant action and change, a relative order that is always in the course of falling apart and being restored, not secure, ambivalent. These oppositions are related to the opposition of knowing/not-knowing. Woodworth would be in the position to *know* about creating homes and families, whereas the narrator states that she is unable to know—about people, or the rules regarding the creation of homes. In a meandering, slightly derailing concluding sentence, the narrator defines the home by outlining an ethics in which she links the impossibility of knowing other people to the imperative to take them seriously: "People we do not know and may never like and can never like, but people all the same who must be considered with the utmost seriousness, the same seriousness with which we consider our own lives" (33). This ethical pronouncement on the need for a considerate attitude toward others links the effort at home-making to discussions of tolerance as indispensable to an inclusive society.

One might repeat the critique that homosexuality, in Kincaid's text, is made to serve once more as the outside factor, the otherness that is needed to create an inside of sorts. Moreover, one might recognize the liberal, heterosexual practice of flaunting one's tolerance of queerness as a way to advertise one's liberalism. However, of the three functions mobilized in my first discussion of homosexuality above (as a point of reference for the stabilizing of heterosexual identity, as a reference to the instability of the signifier, and the evocation of a discourse on nature and the natural that allows one to address basic questions about nature, life, and death in new ways), the last two seem to be much more influential. This, then, is why the reference to homophobia and homosexuality functions as the supplement to the failed narratives of home-making. This reference produces a home that is predicated on the imperative to acknowledge everyone's irreducibly specific way of being, without categorizing this specificity along the familiar social parameters. That is, by acknowledging the instability of such categories. This is not an inclusive definition of the home. Analogy is used at a very basic level only, by placing everyone in the category of "people." There is no other inclusive gesture. There is no defense of other people's equal worth or acceptability; there is no suggestion of empathy, no emancipatory plea. There is no call for an evaluation of any sort, apart from a "consideration." The home thus produced is a practice rather than a space. A practice that does not seek to include, as it refuses the categories that would allow it to create an inside (sexuality, class, gender). Renouncing these categories also means giving up on possession, social legitimacy, knowledge. This text seems to argue that one does not need to be in full possession of an institution (marriage, family, home) to inhabit it. On the contrary, not possessing the institution, not knowing it, allows one to practice the one set of ethics that really creates a home of sorts: a refusal of closure. Though that refusal may only emerge after reducing complex persons to people, organisms, and human relations to a set of rules and negotiations.

This challenge is not intellectual. Readers who are acquainted with Kincaid's work will recognize the same exalted tone as, for example, in *A Small Place*, where the narrator offers another moral reminder about our shared humanity. Such readers will also recognize the utopian quality of such an insight, which seems hard to put into daily practice. If, however, we read this appeal as an effort to outline the most basic, material foundations of an ethical way of being, and connect it to another definition of the home proposed in this chapter, it becomes clear that Kincaid uses the materiality of the body as her point of departure. "If I were asked to make a definition of domestic space, I would say that domestic space is any space in which anyone might feel comfortable expelling any bodily fluid" (23). If the father and daughter agree that the home is a social formation brought about by rules, and if the narrator elsewhere defines the home as an attitude of openness, the home would appear to be a strictly physical place where the boundaries of the body no longer have to be strictly guarded and preserved. This definition embraces several basic bodily functions, including sex, on a physical level where normative discourses of sexual identity are irrelevant. The attitude of openness works on two levels: both on that of the body, of which the inside is opened up, and on that of the larger social body, of which the categories that exclude and include are dismissed. Rather than hinging on an intellectual or ethical pursuit, then, Kincaid's home seems to be defined by a specific embodied experience.

It is at this point that nature comes into Kincaid's narrative of the creation of a home, in which she uses homosexuality as the missing link. The references to gardens and nature are sparse, making this chapter stand out in a book about gardening. However, the former house owner, Woodworth, does not only carry a reference to wood in his name, he was also an academic botanist. Thus, the stability and knowledge that the narrator refuses are related to the botanist's mastery of matters of nature. Nature and gardens are the symbolic possession of others. For example, an expert in landscaping criticizes the narrator's garden design and urges her to get rid of the trees. It is the other characters, not the narrator, who know the stories and details concerning the trees surrounding the house (21–22). The narrator takes her distance from their discourses, apologizes to the trees (22), and continues her own particular musings about her failure to differentiate inside from outside, garden from house, and thus to inhabit the home. For her, this is ultimately a more suitable way of being in the world than the way of the specialists. The references to nature reinforce the emphasis on the need to interact and negotiate with one's relative outside, the material world, without "othering" it, or identifying with it.

What sense of place is created here? No coherent, multiple, inclusive collectivity. A painful, contradictory, incessant practice of connecting and disconnecting, with the help of devices that may or may not be open to an ethical or political critique *and* praise, always opened toward death. Connecting and negotiation are central in this place, but never in a humanist way, since this form of connecting shies away from identification and empathy. A posthumanist place, I would say—which is neither a good nor a bad thing. It shows a deep and bitter insight into the nature of writing in postcolonial times—and into the wayward, vital place of nature within that writing.

NOTES

1. Kincaid has criticized *At the Bottom of the River* for its decency—for example, in an interview with Selwyn Cudjoe, in which she problematizes many efforts to define her work (Hoving, 2001a, 204–6). But also see below for an analysis of Kincaid's opinion of stability and solidity.

2. In *My Garden (Book):*, for example, the narrator expresses her fear of stability: "All three of them are people we would have been afraid of becoming, if only because the stability of such lives remind us of our own youth" (33).

3. See Nadine Dolby (2003), for a lucid account of the colonial insistence of disconnectedness, and the critical ethnographer's effort to create a productive form of connectedness: the tourist (in Kincaid's work) "achieves this freedom through a dynamic of disconnection: by drawing a bright line between 'them' and 'us'" (58).

4. In the same vein, the "purely cognitive or descriptive category" of coherence was "idealized [. . .] transformed [. . .] into a principle of value: to be coherent is to be beautiful, good, desirable, 'true'" (Herbert, 2004, 185).

5. Some critics do so by relating Kincaid to Caribbean theorists of globalization, such as Edouard Glissant and Wilson Harris (Covi). Donatien-Yssa, as I argued in chapter 4, refuses a postcolonial reading of other work by Kincaid to offer a reading based on the Caribbean concept of "la blès."

6. Hume is still a source of inspiration to, for instance, Ian Hacking, who also pleads for sympathy between humans and animals (Lamb, 134).

7. In the wake of Derrida, Bill Brown (2001) differentiates the thing from the object: for Brown, objects become meaningful through certain codes, "because there is a discourse of objectivity that allows us to use them as facts" (4). Things, in contrast, "assert themselves" in their "thingness" on the very moment when they do not function any longer in the expected way. Thus, Brown concludes, "the thing really names less an object than a particular subject-object relation" (4). See chapter 10.

8. Paquet (2002) offers a beautiful analysis of the juxtaposition, in *My Brother*, of Devon's homosexuality and the narrator's authorship (253–54).

9. I suggest that this manipulation is amoral, because I assume that, *for the writer*, during the act of writing, writing is *primarily* (though not exclusively) an affective, or libidinal practice. That means that morality may not be the *writer*'s first concern (though it may be a second

or third). There are more urgent affective needs to attend to. This manipulation is amoral, in spite of the fact that, as an affective or libidinal game, it deals with power. The desires emerging though this practice may very well be analyzed, situated within a social context, and criticized by others (or by the writer once the writing is—momentarily—done).

10. Sagan builds, of course, on Lynn Margulis's work. Margulis is considered the decisive authority when it comes to the theorizing of symbiosis in evolution.

11. In a reading of André Gide, Christopher Lane (1996) ponders the implication for the notion of subjectivity of the acknowledgment that the subject is inhabited by a plurality of contradictory passions that are at war with each other (instead of by smaller organisms, as Margulis and Sagan suggest). This relates to my reading of Scott's novel *Aelred's Sin*.

12. As for example in Bachelard and Heidegger, who function as the main points of reference in a collection of (ecocritically oriented) essays on houses and homes (Smyth and Croft, 2006). Homi Bhabha would argue that, during colonization, the private has been irretrievably invaded by the public sphere, so that home and outside are no longer differentiated.

13. Boyce Davies, 1994, 126; Childs (2006).

14. See *A Small Place* (1988), in which the narrator explains that people "like me" are distrustful of capitalism, because they once were capital themselves (37–38).

15. It is for this reason that Covi (2003), relating to this same chapter, can argue that the "book translates epistemological inquiry into political thinking" (59). I argue that the story does that, but exceeds political discourse too.

16. Building on Marjorie Garber's (1993) argument (see chapter 7) that it is often the transvestite who functions as the supplement, I would argue that homosexuals can also play that part (though in a different way).

17. In the preceding chapter, I argued that Kincaid's text tries to escape the logic of the Imaginary by turning to the Real. Here, we see the effort to escape by moving into the Symbolic, the sphere of laws and institutions. As the text is based on the assumption that the symbolic order cannot grant stability, neither strategy is fully successful.

II

Diversity

Chapter Six

Dwelling

Olive Senior's Assaulted Landscapes

In his 1981 essay *Le discours antillais* (in the 1999 translation *Caribbean Discourse*), Edouard Glissant criticizes the notion of universality as one of the Eurocentric notions that prevent knowledge of the diverse postcolonial world.

> It is difficult to separate theoretically the notion of oppressive dignity from the oppressive reality of private property. This makes sublimation necessary. This explains why Western philosophy and ideology all aim for a generalizing universality. [. . .] (This) is the ultimate weapon in the process of depersonalizing a vulnerable people. The first reaction [. . .] is the stubborn insistence of remaining where you are. (138–39)

"Remaining where you are" would be a good strategy for (formerly) colonized people to resist the oppression that goes with certain forms of theorizing. It is perhaps difficult to understand how "remaining where you are" might help counter the depersonalizing process, as Caribbean writers have, in contrast, often seen the departure into the wider world as a means of gaining knowledge and coming into one's own (see, e.g., Jan Carew, George Lamming). But Glissant goes on to explain what he means by "remaining where you are," which, for him, means staying in the Caribbean:

> What is the Caribbean in fact? A multiple series of relationships [. . .] in the Caribbean each island embodies openness. The dialectic between inside and outside is reflected in the relationship of land and sea. It is only those who are tied to the European continent who see insularity as confining. A Caribbean imagination liberates us from being smothered. (139)

The place where you are is nothing if not a site of convergence of many histories, including colonialism, international political and cultural movements such as pan-Africanism (Marcus Garvey) and négritude. In his earlier work, from which these quotes are taken, Glissant suggests that this particular condition of multirelationality is characteristic of the Caribbean. In his later work, however, he uses the same images and concepts to characterize not only the Caribbean, but the postcolonial, globalizing world in general. For him, every geographical place is a site where other places meet. Therefore it becomes possible to meet the world not just by travelling, but by dwelling too.

These images express Glissant's poetics of Relation, his imagination of the cross-cultural world as a rhizomatic network of relations, in which identities are fluid, open, and temporarily shaped by their relation to others. Significantly, Glissant refers to Caribbean plants and trees to illustrate his model: mangroves, for example, with their entangled roots and branches, and amphibious dwelling, are the counterpart to the strong European oak with its deep roots and unambiguous trunk, which serves as a model for a strong, autonomous identity. Following Deleuze, Glissant differentiates *identité-racine* (root-identity) and *identité-relation*, and with the help of this last concept he invites us to understand identity as the effect of relations, rather than as the cause of relations. In the same vein, places (nations, cities, landscapes) are the complex points of convergence of many cross-cultural and transnational relations. Staying in a place and probing that place thoroughly through analyzing its historical, political, and ecological complexity yields a knowledge of cross-cultural relations and globalization that (literal) travels could never bring. It brings an unparalleled knowledge about the world. On the one hand, it is very much a situated knowledge, for it is knowledge of a very specific place, with its particular conditions; on the other hand, this knowledge embraces the world, as the whole world meets in that single place.[1]

Glissant's work is immensely important, as it allows movement away from a pastoral understanding of place, in which nature is understood as the sign of its authenticity. It also criticizes the association between nonmetropolitan life with harmony and balance, which has shaped antimodern primitivist colonial discourses.[2] The productive reciprocity between the situated and global knowledges brought by "dwelling," however, begins to change in Glissant's later work, when the emphasis shifts to the theorizing of mobility, flexibility, and flight, as the key-terms for a new, more productive narrative of migration. The productivity of such theories of globalization has been seriously questioned. Famously, as seen above, Peter Hallward (2001) has criticized the Deleuzian and Glissantian approaches for their increasing interest in what he calls the *singular* force that energizes globalization (such as Relation), and their inadequate theorizing of the *specificity* of space. Hallward explains specificity as follows: "a specific mode operates, through the active negotiation of relations and the deliberate taking of sides, choices and risks, in a domain and under constraints that are external to these takings" (xii). However, he argues, there would be a growing tendency to conceptualize the postcolonial as a space of difference, in-betweenness, and hybridity. Concepts such as these, which are produced by an essentializing and categorizing colonial discourse, partake in a larger poststructuralist, deconstructive critique of the epistemological violence of binarism. This new postcolonial poetics, however, in spite of its explicit aim to grasp *specificity,* through a critique of binarism, leads to a focus on *singular* forces, that is, the non-relational, autonomous forces that shape the postcolonial world.[3] Hallward mentions among others the later Glissant, Bhabha, Sarduy, and Dib as writers engaged in this literary enterprise. But not all writers and theorists share this singular postcolonial project. Hallward offers a useful critique of the field of postcolonial theory, but he can only do so by acknowledging a *certain* literary canon of postcolonial writing (which he admits to himself, when mentioning, for example, Toni Morrison and Assia Djebar as representatives of another tradition).

Apart from elaborating his critique, one could also offer an alternative historiography of postcolonial writing, and ponder the differences with that more glamorous (but not uncontested) body of fiction and theory that is represented by the predominantly male writers and theorists mentioned above. Such an alternative approach would focus on the women writers Hallward does not mention: Dionne Brand, Michelle Cliff, Shani Mootoo, Ellen Ombre, Marlene Nourbese Philip, Astrid Roemer, and Olive Senior, to sum up just a few of those who would form the heart of another tradition.

Peter Hallward (2001) argues, sarcastically, that "[p]ostcolonial theory often seems to present itself precisely as a sort of general theory of the non-generalisable as such" (xi). He is especially critical of Glissant's interest in totality. After stating that Glissant's critics agree that Glissant is hostile toward universalism and (especially Hegelian) totality (Hallward 70), Hallward goes on to suggest (sarcastically again) that "Glissant is hostile to Hegel's 'totalising Reason' [. . .] only to the degree that it is not totalising enough" (72). Hallward argues that "[t]he essential thing to recognize is that Glissant's Relation is a properly singular concept. (. . .) Relation is a name for self-differentiating reality as such, it is not a 'relationship' between things" (122). For this reason, Glissant does not succeed in opposing universalism.

In contrast, in the writings of that alternative tradition, we find a deep commitment, not to the singular, but to the *specific* (to use Hallward's term)—though it stands to see if, and to what extent, this commitment is translatable into a general theory. For this other tradition does not necessarily bring out a completely different approach to the postcolonial that might be represented as a different paradigm. It does, however, offer a different perspective on the agenda of postcolonial studies. If postcolonial theory is both a "'post-identitarian' *and* context-specific" project, as Peter Hallward has it and Jordan Stouck repeats (66; Stouck, 2005, 110), then this other tradition represents another way to *deal* with the tension between post-identitarianism and contextualizing, by foregrounding the need to relate to a specific place, and to remain there, in that inside space.

What interests me here is the centrality of nature in the description of what makes a place specific. Glissant uses botanical metaphors to illustrate a Caribbean sense of identity (and its self-awareness of being bound up with the world). In this alternative body of writing, too, references to nature function as a means to convey and explore the concrete specificity of a space; not an essential authenticity, but a specificity that is understood as intensely historical and relational. Writers such as Jamaican-Canadian poet and novelist Olive Senior and Surinamese-Dutch writer Astrid Roemer show deep affinity with Glissant's evocation of images of nature as a means to ponder the opacity of place. Like him, they resist the essentialism that often goes with the celebration of nature as the sign of local authenticity. However, they did not follow the later Glissant's development toward singularity. If Glissant is intent on conveying the boundless energy of globalization itself, Senior and Roemer prefer to use natural images to ponder the specific agencies and effects of local corruption and violence—taking the invitation to remain where they are seriously. Their imagination of the environment results in a partly different critique of globalization than their (new) materialist colleagues. Like the early Glissant, they will not be satisfied with merely pointing at epistemological habits such as classification and binary oppositions as the primary cause of (neo)colonial violence, but they focus on the political practice of exploitation itself. They are not interested in phenomenological explorations of the experience of dwelling (as Heidegger is), but primarily in the well-defined political and economic dimensions of globalization, which emerge in the thickness of their specific, local effects. In addition to this first concern, however, these non-metropolitan writers will often also address the isolation that is—paradoxically—caused by the supposedly interconnecting rages of globalization. Writers subvert the stereotypical opinion of isolation as backwardness by showing that this isolation is produced in the interest of an exploitative globalization, and that the notion of isolation is therefore only veiling a destructive connection with the imperialist project of globalization.

In this chapter, I will first discuss how Jamaican writer Olive Senior turns to the Caribbean environment to find the images and patterns that allow her to analyze, to *know*, and to manage the forces and effects of globalization *from within*—from what anthropologist Timothy Ingold (2000) usefully calls a "dwelling perspective."[4] Through a comparison of Senior's work with that of others, notably Michelle Cliff from Jamaica and Astrid Roemer from Surinam, I will argue that a *postcolonial* "dwelling perspective"—which differs from Ingold's definition—represents a different form of knowing postcoloniality and globalization, that does neither join postidentitarian mainstream theories, nor remains steeped in colonial primitivism or nationalist or regionalist realism.[5] References to the environment are crucial to this alternative way of knowing. I will then return to the question what literature can do, faced with social injustice and environmental disaster. I want to show that this writing is not necessarily *merely* concerned with the epistemological debate, not *only* interested in problematizing or redefining identity. A hesitating exploration of the possible affinities between these writings and the field of the environmental humanities (specifically in as far as they are shaped by posthumanist new materialism) allows me to comment on the nature of the writerly project of those who adopt a dwelling perspective.

THE LOCAL, THE GLOBAL, AND THE COSMIC

Olive Senior's poetry, written in the 1980s in Jamaica, and in the 1990s and later mainly in Canada, explores issues of isolation, confinement, and dividedness on the one hand, and exploitation and poverty on the other. The former concern is shared with many other migrant writers, especially from the Caribbean, who, like her, search for ways to reconnect the fragmented, disconnected world (Reiss, 2002). Senior, though, invents a poetic language that links her rethinking of global space to her explicitly political concerns about local exploitation.

A first characteristic of her writing is an insistence to understand space foremost as specific, untranslatable, historical, material. Let us consider this poem from 1985, in *Talking of Trees*. A traveler returns home, and finds

> a landscape redrawn by pygmies
> in your giant's eyes the people shrivelled.
> ("Reaching My Station," 76)

Migration is not only affecting those who leave, but also those who stay—they shrink, and know they are revealed as shrunk. Migrant writers use both perspectives, those of the ones who leave and those of the ones who stay, to address migrant experience. Olive Senior published this poem in 1985, when she was still residing in Jamaica at the age of forty-four, after having spent several years studying in Wales and Canada, when she was in her twenties. But even after migrating to Canada in the early 1990s, she may still choose to write from the place she left—a decision that, in the eyes of many critics, adds to the importance of her work. The experience of migration we encounter here is not a masculinist success story that celebrates the triumph of unlimited movement (as Donnell [2006] argues, responding to Boyce Davies; Ganguly, 1992). What makes this poetry so productive as a countertradition is its interest in the effects of migration on the places where those live who did not migrate. Here, Senior offers a wry comment on the time-space compression that would characterize the postmodern (Harvey, 1989). It concerns not the distance between places, but the desperately poor, assaulted land—and cityscapes of the former colony themselves that have shrunk, whereas the metropole kept its size, as if the time-space compression brutally devours and erases less powerful regions.

> do not say
> it is the sea only
> that divides us
> [. . .]
> How can you in your
> central-heated city know
> how sunspots on unpainted
> tin shacks
> sliver my eyesight
> [. . .]
> earth shifts
>
> new wrinkles
>
> divide us.
> ("Letter From the Lesser World," 47)

In this poem, also from the 1985 collection, Senior offers another critical comment on the assumption of the time-space compression: the distance between the privileged and exploited regions of the world has, in contrast, become almost unbridgeable. On the other hand, even the intimacy of local space is broken open, torn apart, and exploded: "The moon's/been conquered, the hospital/not yet—it's just as far/and alien" ("Reaching My Station," 77).

This description of an exploded homeland reminds one of Homi Bhabha's characterization of colonialism as the invasion of the private space that is then irrevocably turned inside out. But the imaginative tradition in which Senior inserts herself is different. If Bhabha is interested in the boundless productivity of the in-between spaces created by migration and intercultural contact—that is, the force of hybridity and difference itself—and if Deleuze and Guattari, and the later Glissant, are also focusing on comparable, worldwide, (singular) forces of globalization, then Senior joins the tradition that turns to the specific, local spaces that were purposefully corrupted by colonization, but are indispensable dwelling places for the local people nevertheless. Senior, like a number of Caribbean women writers, adopts a "dwelling perspective." In Ingold's words, from a dwelling perspective, "the landscape is constituted as an enduring record of—and testimony to—the lives and works of past generations who have dwelt within it, and in so doing, have left there something of themselves" (1993,152, qtd in MacNaghten and Urry, 1998, 167). For Senior, however, the landscape is not just shaped by those who dwelled there, but also by the political, economic, and ecological invasions from outside. The locals do not *own* their own space. Yet, they continue working it.

For Senior, the dwelling place is not merely forced open to become bare outside; nor is it just a confined vacuum cut off from history. Senior's place comes close to what Lawrence Buell calls the bioregion, a term that implies both geography and consciousness. With reference to Thomashow, Buell (2005) describes bioregionalism as "an attempt to 'integrate ecological and cultural affiliations within the framework of a place-based sensibility'" (2005, 83). The predominant ethic of such a sensibility would be an ethic of sustainability (84). This would be all the more relevant for the modern bioregion, which is marked by a "sense of vulnerability and flux" (88). This vulnerability stems from the awareness of the global context that decides the regional environment: "the locale cannot shut itself off from local forces even if it wanted to" (88). Buell discusses Derek Walcott's *Omeros* as an example of an imagination of place "in multi-scalar terms," embracing both the local to the global (96). Walcott, too, imagines the "vulnerability and porosity of the postcolonial condition" (94). Walcott's interest in the global scale can well be compared to that of other Caribbean thinkers and writers, such as Edouard Glissant and Wilson Harris. But whereas these authors are sometimes more concerned with the global than the local, Olive Senior consistently stays in close contact with the local.

She masterfully manipulates the spatial and temporal scale of her localities to show the very many ways in which they are connected to the globalizing world. Her poetry, set in a sensual local present, may suddenly open, not onto the destructive postcolonial world, but into cosmic space, as when referring to the conquest of the moon above, or on Deep Time—to jump back to the mundane present immediately afterward. Thus, the modest fern, encountered in the neighborhood, is suddenly understood as the *joint* (Stuart Hall, 1980) between Deep Time, industrial revolution, and the ecological disaster of the global present:

> Fern was present
> at Creation fired
> the carbon of
> revolution fuelled
> the age of industry
> in which inheres
> its diminution
> ("Fern," 69)

These sudden poetic flights into Deep Time and Outer Space counter the notion of dwelling places as turned inside out, mere outsides.[6] This poem, published in the 1985 collection *Talking of Trees* and reprinted in *Gardening in the Tropics* as "Fern," also counters the neocolonial notion of dwelling places as relatively uninteresting premodern localities, closed off from the exciting hybridization and acceleration of the larger world. If Senior's vision is radically relational, it is so in the sense of the early Glissant, who focused on political and economic specificity. It is also relational because she situates her localities within cosmic space (and we might recognize an affinity with Wilson Harris here), while simultaneously picturing these localities as knots in an exploitative global network. These two dimensions also characterize her treatment of time.

In the first place, sometimes, we recognize a long perspective, which Macnaghten and Urry (1998) call a "glacial sense of time" (154), as it synchronizes with the slowness of evolutionary time; this perspective ties in with the later discussions around the concept of the Anthropocene (the epoch in which humankind acts as a geological force, influencing ecosystems and the climate) that also call for large-scale imaginations of time (Chakrabarty, 2012; Dimock, 2006).[7]

Secondly, Senior's sharp political statements, focusing on the historical present, form her alternative and more frequently used temporal approach. Senior's appreciation of the world as intensely historical relies on both the evolutionary *and* the political senses of history. Like Glissant, Senior acknowledges the violence that produced, and still marks, dividedness and isolation.[8] Instead of changing the focus to mobility, however, she finds a solution in swiftly alternating between temporal and spatial contextualizations, retaining the small scale of everyday life while zooming out to cosmic scale.

In Senior's swift switches between present and Deep Time, local and global, we can recognize the argument that, in times of globalization, the potentially protective intermediary scale (of the nation, of local communities) has vanished.[9] The vulnerable ecosystems of the South, for example, are no longer protected from direct interventions from abroad. As Spivak says: "The pre-national is now globalized, after an uneven insertion into the nation form of appearance" (1999, 380). Senior's writing testifies to this experience of threatening immediacy, when invoking notions of the "imperceptibly fast" time usually associated with digital speed (Adams in Macnaghten and Urry, 147) and the instantaneousness associated with "postmodern disorganized capitalism" (147) as a way to picture the relations between the Jamaican present and the outside world. Jamaica's local surroundings have become a vulnerable outside. Senior's work addresses the shifting modalities of the marginalized existence in globalization: it responds both to the pain of being mere outside, and the pain of isolation, that is, of being mere inside. Her effort of creating links with a larger outside does not echo the anxious vulnerability to the outside common in postcolonial writing alone, but it represents a triangulation that allows for an escape from the confining present, while refusing a merely political, or merely cosmic ("glacial," spiritual) discourse as an alternative. If Senior links the "here/now" to "glacial time/cosmic space" (the left leg of the triangle), she also creates a link between the "here/now" and the political/economic field of "imperial history/globalized space,"[10] a field that is in turn linked to "glacial time/cosmic space," thereby closing the triangle. The triangle could be expressed in figure 6.1:

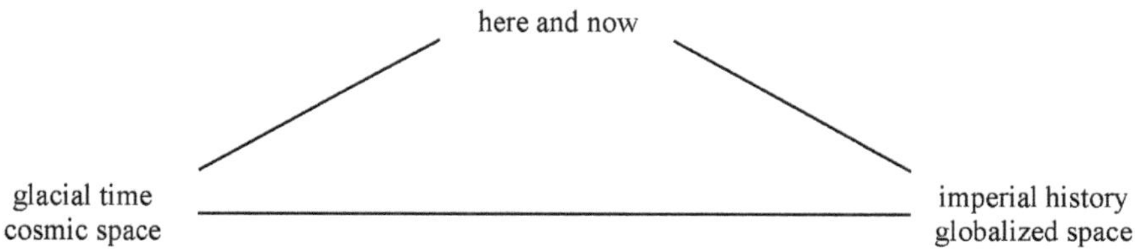

Figure 6.1.

The link between the present and the political powers of imperialism (the right leg of the triangle) is clear enough in Senior's very many direct references to exploitative leaders, crime, political violence, and poverty. The fern poem is one example of the way in which Senior's poetry produces the necessary link between the political field and glacial time. It explains industrialization, exploitation, and corruption as inherent to nature itself. Thus, she avoids a reductive scheme of neocolonial, technological evil-doers versus local, innocent nature-dwellers. Nature is never innocent. Nor is evolution "good." Remarkably, in Senior's imagination of the globalizing world, it is the level of global space itself which often disappears, not that of the nation or region. The global is understood either as the cosmic level, or as the global political strategies of exploitation that always materialize in their local, specific forms. True, Senior addresses the global level when acknowledging the influence of metropolitan spaces on her Jamaican dwelling space, but she addresses the metropoles always in their specific, parochial nature, as isolated localities; for these "central-heated" metropolitan spaces are not able to understand what happens elsewhere on the globe, as in the poem above. In her triangulating moves, she never loses the specific materiality of place out of sight.

This literary strategy, in which the cosmic scale plays a crucial role, calls for the question to what extent the contemporary critical interest in the Anthropocene could serve as a frame for the reading of Senior's work. Indeed, Caribbean literature offers a wealth of references to geological scale too, but in a very specific way—they are often related to the anticolonial search for origins and identity. In their groundbreaking study on Caribbean literature and the environment, DeLoughrey, Gosson, and Handley (2005) explore why Caribbean writers have been fascinated by the question of the origins of their geography. The reasons lie in the fact that, in the Caribbean, "geography serve(s) as the metaphor for history" as Gordon Rohlehr has it (1981, qtd. in DeLoughrey et al., 3). Caribbean history has been erased for a number of reasons; the erasure of indigenous peoples by the advent of European colonizers, the destruction of local ecosystems by the plantation system, the absence of precolonial historical sources, and so forth. Even the notion of a Caribbean history itself was erased, as it was overwritten with the European myth of a timeless tropical Eden (12). Caribbean writers have therefore insisted on the importance of the land and sea of the archipelago themselves, as the source of knowledge on the layered and complex Caribbean cultural identity and history.[11, 12]

Geography and geology can act as archives because the Caribbean is, in the first place, a crossroads of global connections—as a repeating island, or a global meta-machine, that is, the engine of the triangular trade; or as a cross-cultural, cosmic junction. Caribbean space is by definition global, cross-cultural, and it can therefore only be approached through a connection with the material global networks that have called it into being. This is the context of the fascination with the global, the cosmic, the primordial, and the non-generalizable universal we find throughout Caribbean writing, from Antonio Benitez-Rojo to Edouard Glissant and Wilson Harris. While this shared imaginative project was originally motivated by the need to ground a Caribbean cultural identity, it is also explicitly exploring the potentials of a post-humanist, materialist approach.

In this writing, the primordial and the cosmic (as accessible through the material landscape) serve as the place where creativity is unleashed, as a hotbed for the imagination. If the Anthropocene is addressed, it is less with the purpose to create an alternative scientific foundation for the rethinking of the Caribbean (as in much metropolitan writing) than because it sparks an alternative poetic imagination. The environment—a crossroads of global, cosmic, geological, economical, and political forces—even functions as a space of visionary potential. Guyanese writer Wilson Harris (1983) is perhaps the Caribbean writer who rejects anthropocentrism most strongly. He locates the necessary large-scale creative imagination in what he calls the cross-cultural "womb of space": "the environment is pregnant with endless evolutionary possibilities" (87, qtd. in Maes-Jelinek, 2005, 251), he says, and, fascinated with "unfathomable wholeness," he speaks of "a living cosmos . . . that may appear to sleep . . . but is susceptible to riddling proportions of eruptive life" (252).

Nevertheless, one of the distinctive qualities of the Caribbean geological imagination is that it is closely, and explicitly, intertwined with a social if not political postcolonial project. In this way, it counters a problematic aspect of the radically posthumanist imagination to humankind as a geological force. In a recent essay, Teresa Shewry (2014) lucidly explains what is at stake. She does so by exploring the cultural and political significance of references to the ancient continent of Gondwanaland by writers from Australia, New Zealand, and India. These references may support different political agendas. A writer from the settler community, for example, would focus on geological developments to avoid more contemporary explanations for the deplorable state of the environment. Shewry summarizes: "As critics have noted, deep time perspectives on environmental and social change may in this way erase human actions, such as the material practices and ecological transformations instituted under colonialism and capitalism" (270).

This useful remark might well explain the context of Senior's writing. A single-minded focus on geological scale might well be a writer's way to avoid political analysis; for some writers, the modest, everyday interactions between people and the environment are worth as much attention, when they try to understand what the Anthropocene feels like in the flesh. Shewry's analysis resonates with Chakrabarty's challenge to explore human agency on multiple scales, and not restrict oneself to the seductions of fathomless Deep Time alone. The postcolonial reminder that all theoretical frames are linked to political agendas is an indispensable insight for the shared exploration of the Anthropocene.

THE LEPERS IN THE FOREST

The strategy of articulating a connection with the larger world without dismissing the specific, local significance of place is a characteristic of a broader literary tradition. This focus can be found in the work of many other Caribbean women writers, even when they have moved away from their place of birth. They are not merely writing about the desire to reconnect with the landscape of their youth, but try to write their stories *from* the experience of the locality that was their home. Moreover, they are fully aware of the fact that their former dwelling space is in deep crisis, and they refuse to romanticize it. In this section, I will first explore this last, painful characteristic of this tradition.

The two first novels by Jamaican writer Michelle Cliff testify to the agonized desire to be part of Jamaica's politically corrupted society, however much it is tainted by a forbidding colorism and heterosexism, and imagine in what ways the Jamaican landscape itself could allow for a return. In the Dutch-speaking Caribbean, *grande dame* of Surinamese literature Astrid Roemer[13] as well as Surinamese-Dutch poet and writer Cándani demonstrate the need to inhabit, however imperfectly, the destroyed space one once called home, and wants to call home again. These writers diagnose the social crisis in the Caribbean primarily as an effect of its isolation, divisions, and failed responsibilities. In these works, the natural landscape is evoked to mediate the reconnection, even if it is a broken, corrupted landscape.

Taking Aimé Césaire's heritage seriously, these writers stay true to his insight that the motherland—the assaulted Caribbean landscape—is irretrievably corrupted, but nevertheless the place where reconnection must take place. Astrid Roemer, for example, organizes her important postcolonial trilogy about a Surinamese family that is divided by migration, corruption, and murder (1996–1998) around the need to acknowledge their intimate connection with corruption. The trilogy is not primarily about the postcolonial fragmentation of the Surinamese community, of which one half migrated to the Netherlands; rather, it addresses the murders by Surinam's postcolonial military regime (1982), and reflects on the way in which the nation (including the emigrants) can come to terms with social and political corruption. The last part of the trilogy, *Was Getekend* (Was Signed/Was Marked), offers a solution. One of the main characters is a foundling, who was already affected by leprosy when he was abandoned, his parents being lepers too. An allegorical reading has no trouble recognizing his leprosy as signifying Surinam's social corruption.

Together with the theme of fire, the theme of leprosy runs through the narrative, where it resonates with metaphors of corruption and rot: they refer to the political corruption that lead to the historical murders in December 1982, but also to the many abortions used to erase the illicit sexual liaisons of the Surinamese elite. On the other hand, leprosy is associated with the Surinamese jungle, where the lepers were banished, and, as so often, the forest stands for the nation's origins, its authenticity, and therefore a precolonial Surinamese identity too. The novel suggests that there has never been an untainted past. Corruption is part of nature. In a sensual play with associations, the novel establishes a web of connections between leprosy, jungle and wood, between corruption and authenticity. The lepers are compared to melting chocolate (evoking an ambivalent historical imagination of colored skin), and (wet) wood; a leprous man is indicated as the Wooden Man. The metaphor of wood creates a web of connotations: it stands for family and authenticity, as the protagonist's father was a woodworker, and the protagonist himself, a gardener, feels most in touch with himself when he is building a wood fire. As the wood is transported from the very heart of the jungle to civilization, it can also be seen as mediating between primordial Surinameseness and the postcolonial town. When used to refer to a state of mind, wet wood has a negative connotation. But wet wood also signifies the floating buoys in the ocean between Surinam and the Netherlands, and with that, it functions again as the mediator between intimately but uneasily related spaces. With the help of these botanical metaphors for connectedness (wood-leprosy), Roemer's narrative proposes a complex fictional solution to the postcolonial corruption. On the one hand, she envisions reconciliation: the healthy protagonists and the lepers engage in a lasting friendship. On the level of society, a public tribunal is established, comparable to the South-African Truth and Reconciliation Commissions, so that the involvement with the murders and the abortions can be made public, shared, and worked through. On the other hand, the mixed metaphor of wood, fire, and corruption is elaborated in a cleansing ritual that cannot be motivated from a psychological or political perspective, but which does close off a painful past. With reference to a historical plane crash in Surinam in 1989, the novel ends with the death of the protagonist's loving white foster mother and his best friend, when their plane collides with a magnificent tree near Surinam's main airport. This crash puts a metaphorical end to the intimate colonial relations between Surinam and the Netherlands. This ending, however, does not undo the trilogy's message that certain forms of corruption and ambivalence are an intrinsic part of postcolonial society, and must be acknowledged—not dismissed or erased, but managed.

In Michelle Cliff's third novel, *Free Enterprise* (1993), one of the heroines also finds herself in the company of lepers. This community, living on the banks of the Mississippi River, allows Cliff to theorize the way in which racism prevents a society to know itself, and thus to overcome its destructive fragmentation. Implicitly building on McClintock's work on the colonial obsession with hygiene and contamination, Cliff lays bare the connection between leprosy and racism, by describing how leprosy was assumed to be a foreign disease, deliberately imported into the United States to contaminate the country. Africans were thought to be carriers (Agosto, 1999, 125). Leprosy is thus associated with the paranoid, violent manner in which racist U.S. society deals with otherness. Cliff counters this racist exclusionism. She does not depict the lepers as an isolated community, destined to die, but, instead, she shows that the lepers, who come from all parts of the world (e.g., Hawai'i, Tahiti, Spain, and Surinam) are the lively carriers of a broad, transnational oral history that subverts dominant historiography (Agosto, 42; Chancy,1993, 268–69; Johnson, 2009, 128). Cliff's heroine (a Jamaican light-skinned abolitionist, who, subject to a discourse comparable to the U.S. attitude toward lepers, was seen as contaminated by the "black blood" in her family history) understands that one cannot be fully part of a multiple, inclusive history if one doesn't embrace contamination. Therefore, she socializes with her leprous neighbors. As Erica L. Johnson points out, Cliff's leprous storytellers do not gather in a "'safe' space far from plague-ridden Florence," as Boccaccio's characters did in the fourteenth century, but they find out "that in contamination lies knowledge, mutual understanding, and self-recognition" (128). Indeed, Cliff understands that interaction between human beings will always bring contamination—physical, affective, intellectual.[14] Like Roemer, then, Cliff argues that a postcolonial community can only thrive when it accepts its own corruption. Knowledge of one's society, of one's self, is only possible when corruption and contamination are recognized as inevitable—they are the unavoidable by-products of contact.

Both Roemer and Cliff suggest that this deeply seated colonial discursive constellation cannot simply be undone. Instead of dismissing the notions of corruption and contamination, we would do better to embrace them. It is here that we see how references to the environment help to build the social and political argument: the literary argument that corruption is an inevitable part of nature helps to bring the point home.

Jamaican writer and critic Michelle Cliff has been positioned as a migrant writer, who articulated the wish to belong from an outside position.[15] Though I contest that interpretation, it makes sense to read Cliff in relation to Senior, and wonder why Senior was more often seen as someone who wrote from an insider's perspective, even after she had already immigrated to Canada. What does this contrast in reception tell us about the differences between a dwelling perspective and a migrant perspective? Their understanding of corruption is very similar, but there are marked differences too, and these will help me to highlight the specificity of a dwelling perspective. Let us first see that for Senior, too, the land's corruption, that is, its unstoppable tendency to creolize and intertwine, is a fact of life. Moreover, it is the very reason why connectedness is inevitable:

> Gardening in the Tropics,
> you'll find things that don't
> belong together often intertwine
> all mixed up in this amazing fecundity.
> We grow as convoluted as the vine.
> ("The Knot Garden," 88)

In this wonderfully ironic poem, "The Knot Garden," from the 1994 collection *Gardening in the Tropics*, the notion of corruption is here understood as the tendency to creolize. This fragment thematizes not just creolization, but also the postcolonial leadership's frustrated desire for control, and its denial to take responsibility. The Leader is the cosmopolitan air traveller, who does not want to land, as he cannot deal with Caribbean creole chaos anymore.[16] The remarkable shift in these sentences is that from "things" to "we." The intensely amoral and unpredictable creolization of this convoluted nature implies human beings and culture. Nature itself is already corrupted, convoluted, twisted; it was not colonization that corrupted it, corruption was already part of its nature. Another poem in *Gardening in the Tropics*:

> Plants are deceptive. You see them there
> looking as if once rooted they know
> their places; not like animals, like us
> always running around, leaving traces.
>
> Yet from the way they breed (excuse me!)

and twine, from their exhibitionist
and rather prolific nature, we must infer
a sinister not to say imperialistic

grand design.
("Plants," 63)

Colonization may have been guilty of submitting, exploiting and destroying nature, but nature itself is also imperialistic. Or, as a traditional gardener warns in another poem in the same collection: the "world out there is full of hellish/creatures" ("Advice and Devices," 111), such as harmful insects.

But the most hellish creatures are human beings. The successful gardener takes all these threats in one stride, by applying the right horticultural and magical methods. Life is full of threats. Colonialism, postcolonial violence, and poverty are but some of the harms life will inevitably bring. Senior's postcolonial dwelling perspective is characterized by her focus on the inhabitants of the corrupted landscape, and their discursive and material strategies to counter a historical variety of threats. As she focuses on the local strategies of survival, she defines colonization and globalization as specific, local forms of violence and destruction, as that is the way in which they materialize in the Caribbean. Senior does not analyze globalization as a completely new spatio-temporal phenomenon, distinct from imperialism, as, say, Michael Hardt and Antonio Negri do. Globalization's material dimensions are as oppressive as those of imperialism, and it does not offer those in the South any more opportunities for resistance and creativity than earlier forms of imperialism did.[17] Senior would agree that globalization consists in direct invasions of transnational institutions into the local, without any intermediary national level to offer protection (the national has turned into exploitative cosmopolitan elite), but, for the less privileged people in the South, that is not a new situation. There is a historical continuity between their experiences of all colonial and imperialist forms of violence. Senior's imagination of the Caribbean environment as intrinsically colonizing, deceptive, corrupted, and destructive allows her to convey that age-old historical experience. What we can say so far, is that Senior departs from the assumption that the world *is* already connected, however violently, and that the problem lies in the way this connection should be *managed*. Cliff rather focuses on the colonial (and nationalist) denial and *fear* of connectedness; for her, the problem lies in reimagining the world's connectedness.

DIGGING UP BONES

Senior and Cliff both see violence at the heart of postcolonial connections. But that shared insight leads to very different imaginations of nature. Senior's acknowledgment of the death that lies at the heart of the Caribbean environment becomes very clear in the poem "Brief Lives," the first of the cycle "Gardening in the Tropics" in the collection with the same title: "Gardening in the Tropics, you never know/what you'll turn up. Quite often, bones" ("Brief Lives," 85). These may be the bones of the "desaparecidos," political activists and dissidents who were killed by a dictatorial regime, or of a young man who just got lost in inimical territory. Remarkably, the gardener who digs up the bones then decides to bury them again, in the expectation that this will allow them to keep growing ("Brief Lives," 85). Senior speaks about the potential of growth and expansion that death offers—as death mediates the return to nature's materiality (see also Cándani, chapter 7). But Senior does not see that promise as certain: a gardener can expect to suddenly dig up human bones of those killed in political violence in her garden, but will then have to *act* to transform that death into growth. Here we discern a nontranscendental, local response to post- or neocolonial violence: it is intensely pragmatic and oriented toward the future.

The image of human bones, buried deep in the soil, when it occurs in other Caribbean writing, may elicit another response. In Haitian writer Edwidge Danticat's *The Farming of Bones* (1998), the expression "farming of bones" refers to the hard work in the cane fields, which shreds the skin and therefore brings the bones closer to the surface. The expression also refers to the unearthing of the traumas of the past by which one is haunted. Furthermore, the expression refers to the novel's effort to resurrect the silenced story of the 1937 massacre lead by Dominican dictator Trujillo against tens of thousands Haitians living in the Dominican Republic. For Danticat, writing consists in the painful digging up of bones. As such, it is an antidote to nature, which "has no memory" (Danticat, 309).

For Michelle Cliff, in contrast, nature *does* have memory. Her oeuvre as a whole is dedicated to the unearthing of erased knowledge of colonization, slavery, and globalization. Importantly, Cliff insists that this knowledge is contained by bodies, material objects, and landscapes: Cliff emphasizes the "materiality of memory" (Johnson, 135). The erased memories of the silenced victims of a racist history are therefore not only retrieved by storytelling. The Jamaican landscape itself contains those memories. Cliff's long fascination with Morrison's statement that "everything is now" (Cliff, 1994; 2009) is an apt summary of the postcolonial shift from history and genealogy ("everything") to space and landscape ("now"): the past is still present; the landscape and the very earth, plants, and trees around us embody it.[18, 19] This significant, historical landscape, however, is presented as highly ambivalent. If the revolutionary daughter in Cliff's *No Telephone to Heaven* (1987) hopes to return to the island where she was born by claiming her deceased grandmother's estate, she finds that an easy reconnection is out of the question. The once fertile, motherly lands have "ruinated"—turned to wilderness when left alone. That reversal could be read positively, as the land's "attempts to return to itself, to reclaim its past" as Knutson (1996) proposes (283), as "the common decolonizing trope of landscape is the unceasing movement back to an original, pre-colonial, or primordial state" (283). But Cliff shows that this primordial state is, before everything else, a state of uncontrollable decay. Her anxious imagination of the native soil is close to Senior's, in that she also dwells on its boundless, rampant fertility. It would be so excessive, that "the fertility of the island is compromised by too much growth," as the former well-ordered agricultural plot lapsed back into wilderness (Smith, 158). The former grand house of her grandmother's estate is experienced as both a womb and a grave, a dark humid space of spider webs, rat droppings, and old shells that refer to the island's origins (Smith, 157; Knutson, 294). The soil gives the heroine a sudden access to glacial time, as Knutson argues, when comparing Cliff's work with that of Wilson Harris: "Harris, like Cliff, recognizes the significance of landscape to provide the catalyst for the 'sudden eruption of consciousness' which explodes in a plethora of past and primordial associations" (279). Both Senior and Cliff combine the function of landscape to evoke a glacial consciousness, with its function to contain a more recent social and political past. But in Cliff's novel, this glacial consciousness is related to death.

Soon after, the heroine is killed during a failed revolutionary action. Her bones are then united with those of her foremothers, who were buried in the very soil she could not really return to when still alive. In the materiality of death, differences once inscribed on the body, such as color, class, gender, and sexuality, can no longer separate her from her family. The bones open the possibility to reconnect with the land by dying, and thus joining them, rather than bringing these bones to the surface as a strategy of survival after trauma (as in Danticat's novel). Thus, we can indeed say that the Jamaican soil, fertilized by bones, forms an alternative history. But these buried bones have another significance altogether than in Danticat's writing (Agosto, 38).

Cliff is often praised as a writer who creates a polyphonic counterhistory of globalization through retrieving the silenced maternal voices. This counterhistory would allow all those excluded from hegemonic discourses to reinscribe themselves into world history. As Jennifer J. Smith argues, Cliff's work should nevertheless not be read as a celebration of a daughter's reinscription in a matrilineal genealogy (which is a widespread narrative in Caribbean and African American women's writing and criticism), but rather as a critique of the idealistic rewriting of history as a maternal narrative.[20] Her novel questions the productivity of the figure of the grandmother (and especially its connections to marronage, through Nanny of the Maroons, and the Jamaican landscape) as a model for resistance and knowledge, at the very instant when it proposes it (Smith, 2009). The exclusionist nationalist and feminist discourses that offer Jamaican nature, the maternal landscape, as the paradigm for postcolonial identification are only empowering to those who can adapt to the role of the dutiful, obeying, heterosexual daughter. But for Cliff's heroine, who is physically unable to have children, the fertile, maternal nature cannot offer a context for identification, so long as it is seen as synonymous with procreation (see Greta Gaard in chapter 2; but also see Bagemihl in chapter 8). Cliff explores the *soil*—nature's materiality—instead of the *landscape*—defined as fertility and procreation—as an alternative frame for identification. This manner of reconnecting to the homeland comes close to Kincaid's narrative in *My Brother*: it is only accomplished by a return to bare materiality. Nature, as the domain of living bodies, in which the categories of exclusion are inevitably inscribed, and with its strong association with the destructive maternal, cannot offer a counterdiscourse. Only death can.

Olive Senior's poetry has a different outlook on the bones that are buried in the earth, and on the function of the soil as a ground for identification. Remarkably, Senior's poems tell us the silenced history represented by these bones almost casually, as if it is already known. Perhaps it is: Senior presents these histories of murderous violence as the very fabric of life in the tropics. The gardening of bones cannot be read as a metaphor for individual return or reconnection to a traumatic past. I do not just want to explain the absence of the metaphor for a desired return as a result of the fact that Senior, while writing her earlier work, was still living in Jamaica. The violent past she refers to here was at that time no more available to Senior than it was to Cliff. First, her mention of the *desaparecidos*, which refers to Latin-America, suggests that she evokes a transnational history of oppression: the first line ("gardening in the tropics") also points at a transnational orientation. Second, such histories are for everyone, dwellers and emigrants, hidden histories of denied state violence and unreported killings.[21]

My point is that Senior does not use the trope of the bones as a possible point for individual reconnection with a hidden past, but as a means to reflect on the *manner* of one's connection to a larger, global network of oppression of which one is an unwilling part anyway. Stuart Hall's notion of articulation will help to explain her literary strategy here, as it also refers to the possibility of creating a provisional coherence from disparate, disjointed elements.[22]

The buried bones occur in the first, the third, and the seventh poems of a cycle of twelve poems, entitled "Gardening in the Tropics," published in the 1994 collection with the same title. Throughout these twelve poems, we find a counterhistory of New World violence since the European invasion and colonization. It is told in a colloquial style from the perspective of the ordinary people who were suddenly confronted with lethal colonial madness on a daily basis. The cycle narrates many strategies of resistance, ranging from the questioning of enforced religious habits, marronage, the evocation of indigenous knowledge, the quick strategies of cultivating food in hidden spots in the jungle in the days of colonization and slavery. The poems also discuss tactics that are useful nowadays: from evading and dismissing governmental control and tuning in with the many underground tactics of resistance that are developed in everyday life, to the uneasy adaptation to exploitation—which is always seen as global.[23]

The first poem ("Brief Lives") insists that the bones of those killed by criminal violence should be buried again; the seventh poem mentions bones as the embodiment of a critique of environmental destruction that masks as Enlightenment. When deforestation has run its course and the jungle has been cut down, "only our bones/will remain as testament to this effort to bring/light" ("Seeing the Light," 95). These references to the bones as counternarrative resonate with the suggestion of the uncontrollable vitality of underground life elsewhere. We find this suggestion very explicitly in the fourth poem, from which I quoted above. This poem discusses the "knot garden," and refers to the hidden fertility of underground life: "so many things/propagate underground, by/division" ("The Knot Garden," 88). Bones are a vital part of that underground fertility. As signs for local and global criminal (governmental) violence, they are active components of an underground counterhistory, as they have been since the early inhabitants of the Caribbean made flutes out of the bones of vanquished invaders.

One could understand these bones as mere absences, as death, as the site where the global destroyed the local, and as a site of difference that was turned into a site of destruction, erasure, and forgetting—a blank space, a gap. However, following Hall's lead, Senior understands the gaps in the torn network as potential *joints*—that is, as the sites where connections can be articulated at the precise point where their difference is also manifest. Instead of using Hall's concept, though, Senior prefers the concept of the *knot*: that creolized vegetable joint that is uncontrollable, but fertile, and indispensable to the garden where food is grown. These knots emerge in the interface of globalization and the local, showing that this joining is violent, resisted. The buried bones, too, are *knots*, rather than joints. The advantage of this metaphor is as follows: First, that it shows that the world is already intensely connecting of itself and in itself (not harmoniously, but through invasions, killings, and resistance). Second, that it highlights the violence in the way in which the postcolonial world is linked—the bones testify to the criminal nature of the money-obsessed postcolonial regimes and gangs, that cannot be said to peacefully *join* the local population, as the locals are violently entangled in the torturing knots of their lethal networks. Third, the wayward notion of the knot is more explicitly situated in a task space, thus moving attention away from the noun (space, joint) to the verb (to knot, to intertwine, to entangle, to manage, to garden, to struggle).

This insight in the deadly entangling of globalized nature, also leads to an anxious understanding of writing as a hazardous venture (see also deCaires Narain, 343). "Do not/write/life poems/the worm/will/hear you" (32), the poem "Nature Studies I" in *Talking of Trees* warns. Nature is treacherous; it hides and deceives, patiently waiting its turn. The poet, inspired by nature to write a life poem, gets entangled with death instead of life: "The poet's lover/contemplates children/the poet/studies suicide" ("Nature Studies I," 33). Nature finally takes over, by forcing its own rules on the poet ("the poet writes:/how weep/when trees conspire/to align me/vertical/and all the time/mountains?" ["Nature Studies I," 34]). Apparently, nature does not allow for melancholy. Writing will evoke nature's devouring vitality, in which both poetry and nature itself are destroyed: "the poet writes/a life poem/the poem self-destructs/the worm/devours/the mountain" ("Nature Studies I," 34)). This poem seems to argue that writing can never be a metalanguage. It is animated by the same forces as nature itself—including the destructive ones. However, the text by Martin Carter which provides the epigraph to the collection as a whole, advises writers that there is discontinuity between—and within—all living organisms. One can try and communicate with trees, but "what the leaves hear/is not what the roots ask." Art, human life, and nature are entangled, but they are transparent neither to each other nor themselves. Senior's imagination of globalization, then, is not aiming at a celebration of connectedness, as if that were the remedy for globalization's ill effects. It is rather a continuous search for the best tactics to survive a particular assault. In this sense, Senior is refusing the transcendence of that postcolonial literature that aims for the radically singular, as Hallward (2001) would have it (16), and aligns herself with the writers who wish to understand the specific instead.

Some work by Senior has been read as evoking a context of death (Maisonnat, 2001), and poems such as "Nature Studies I" do suggest that an intimate encounter with life's vitality itself may be killing. In that aspect, Senior's work comes remarkably close to Cliff's. Indeed, I can now point out the affinities and differences between their imagination of globalization through the theme of nature. For both, the world's connectedness, or creolization, is a crucial insight. Both writers are politically committed, both write about outsiders and marginalized.

Senior accepts the world as nothing but an incredibly vital, violent, and therefore also deadly process of creolization, that keeps intertwining incommensurable, untranslatable events and sites. Her writing is meant to negotiate rage in such a way that survival becomes possible. Cliff primarily explores the colonial and nationalist dismissal of creolization, and focuses on the exclusionary effects of that denial. For her, creolization can only become manifest in the margins; through vodun magic, in communities of outcasts, and in death. These two projects are complementary. The main difference between Senior and Cliff, then, lies in Senior's pragmatism. Senior's pragmatic voice is marked by the urge for survival that is tied in with the comedy (Meeker, in O'Brien, 2001, 150–53). Though I do not think Senior's work should quite be read lightly, her voice is indeed pragmatic, ironic rather than cynical, and opened toward the future. The following section will focus on this pragmatism and the crucial notion of agency in her poetics.

FROM POSTCOLONIAL SPACE TO TRANSNATIONAL TASKSCAPE

To what extent can we now say that Senior's writing embodies a dwelling perspective? The first characteristic we encountered was Senior's insistence that *dwelling is a practice*. She changes the question, how to imagine global space, into how to *act* in a corrupted landscape? She approaches space not as landscape, but as a difficult garden to be managed, or a stretch of land that has to be worked hard.

In *Poetics of Relation*, Edouard Glissant (1990, 1997) proposes to describe the destroyed landscape of the postcolonial Caribbean not as environment, but surroundings, in order to indicate the loss of the former social and cultural meanings it received by the communal labor that shaped it (151). Senior, and other Caribbean women writers, such as Kincaid, Cliff, and Roemer, do not dismiss this analysis, but they do not restrict themselves to mourning the loss, and explore instead how that assaulted landscape is still labored. In their work, global space consists of a series of ambivalently linked (and disjunct) historical and social local spaces that take shape through difficult human labor, while human lives are shaped by that locality too. With a felicitous term, anthropologist Timothy Ingold (2000) proposes to call such spaces taskscape—which is a less ambiguous instance of the lived space that geographer Edward Soja differentiates from real space or imagined space than landscape (Macnaghten and Urry, 167). Importantly, if Senior talks about gardening, she doesn't refer to the visually pleasing flower garden, but to the functional vegetable plot, the follow-up of the earlier slave gardens, which were a means of survival.

This emphasis on local agency, however, is not enough to support the argument that Senior's perspective is tied to its Jamaican locality. Jamaica Kincaid, for example, also offers an approach of space as taskspace (see chapter 4), but she doesn't offer a dwelling perspective. Another argument that supports my statement that Senior's work embodies a dwelling perspective is that we find here *a culturally specific* experience of the landscape. This perception of the land could be called non- or even premodern, as it takes its distance from the modern visual approach to nature as landscape. Anthropologist Kay Milton (1993) made a useful distinction between "land" and "landscape," paraphrased by MacNaghten and Urry in this way: "The former entails conceptualising land as a physical, tangible resource which can be ploughed, sown, grazed, built upon. It is a place of work, conceived of functionally rather than aesthetically" (200). To make a comparable point about alternative understandings of the land, Canadian critic Susie O'Brien differentiates a botanical view from an agricultural view. In relation to Kincaid's work, she points out that Caribbean conceptions of nature may not be called "green," or environmentalist in a narrow sense of the word, as Kincaid's work suggests, as these conceptions diverge from an American sacramental view of the wilderness, which could be called "liberal environmentalism" (2002, 173). Dutch cultural theorist Petra Halkes refers to environmentalist Neil Evernden to differentiate a Romantic notion of nature from a medieval one, when nature was not seen as following "'natural' laws, but is as unpredictable and capricious as humans. [. . .] Knowledge of nature is *emphatic*, since it relies on 'an underlying similarity between the human and the natural world'"(21).

These definitions help me to describe Senior's approach more accurately. Steering free from the modern perception of the garden as a visual space, she evokes a local/global space that is brought into being by the incessant labor of both human beings and nature's own fertility and creolization. At this point, the nature of Senior's effort to know the global world through interaction becomes clear. Her poetry doesn't "know" the world because a supposed "underlying similarity" between the human and the natural world would make each transparent, accessible to each other. Rather, she suggests an irresolvable intertwining; the analogy between nature and humanity lies in the fact that they are both "unpredictable and capricious," but their conditions are incommensurable. Instead of a knowledge based on analogy and transparence, she pursues a difficult form of experimental interacting, based on oral wisdom, as the means to know, or rather: relate to nature. This perception is less premodern than *critical* of modernity, while leaning on premodern indigenous knowledge. As Denise deCaires Narain (1996) states: "Senior insists that her major cultural influences came out of the oral culture of rural Jamaica: preaching, praying, testifying, concerts, and storytelling sessions" (341).

This brings me to the third way in which we can describe Senior's attitude to nature as local. Her attitude cannot be disengaged from *local speech*. Senior's flexible pragmatic, ironic voice, moves between Standard English and creole, a language which testifies to the "tensions and contradictions that structure our engagements with the physical world" (181), as Susie O'Brien says (2002, 180). Language structures one's engagement with the world; but the environment has also shaped the structure of language. Kamau Brathwaite (1984) is, of course, the most notable theorist of this intertwining. In *A History of the Voice,* he outlines the need to develop (on the basis of creole and oral genres) a nation language that would be in tune with the rhythms of the Caribbean. Glissant, too, insists that "landscape has its language—words are entangled [. . .] in the mobile structures of one's own landscape" (1989,

146). Alfred Hornung (1998) even speaks of the "langscape" in Cliff's writing.

Environmental theory helps us to recognize the different relations between human beings and nature that are expressed in the different languages and discourses in the world, and the different forms of human agency. The writers of *Greenspeak* discern—rather obviously, perhaps—three main patterns: Humans may be seen as controlling nature, nature may be seen as controlling humans, or humans may be seen as a part of nature (Harré, Brockmeier and Mühlhäusler, 1999, 140). Such perceptions are consolidated in language, they argue; most European languages, and new national languages such as Indonesian, testify to the first perception, as can be judged from the dominance of causative words, which make humans the prototypical agents of activities such as killing, curing, planting, and so forth. Other languages do not show the same centrality of this kind of causality, and allow agency to other entities than men or animals (140). Senior's poetry has been praised for its subtle, accomplished use of the many linguistic registers of Caribbean creole, from "proper English" to deep creole (Pollard, 1988; deCaires Narain, 1996; Patteson, 1993). Strikingly, her use of creole assigns agency to trees as much as to human beings: "Jerusalem Thorn crown me/Almond nurture me/Guango shade me" (84)—a grammatical construction facilitated by creole speech habits. That doesn't mean, however, that there would be an equal, harmonious relationship between human beings and their taskscapes. Nor does it mean that the world can be imagined as a series of organic or inorganic machines, or relations. Instead of an analogy, a binary relation, or a network of relations Senior assumes much more complex, violent forms of interactivity, that are specific to their localities, and that are decisively humanist (Stouck, 2005, 104, 111).

Postcolonial theory has always been interested in issues of identity, especially when it let itself be inspired by psychoanalysis. Nowadays, the most popular competing inquiries into issues of globalization and nomadism are, in contrast, anti-identitarian. Elizabeth Grosz (1994) differentiates this new materialist approach by explaining that it would be interested in the body, rather than identity, and less in the experience of migration than in the question how the body acts, as a social object, that is, how it connects to its environment. Though I want to differentiate the tradition I am trying to highlight here from new materialism, it does share this focus on the body as an acting, connecting organism, but in a specific way.

To a certain extent, Senior's work can be said to foreshadow a posthumanist approach, as articulated, for example, by feminist philosopher Karen Barad. Barad suggests that knowledge about the capricious world (that includes the human and the nonhuman) is not based on, but *consists in* one's interaction with it, or, to put it more precisely, consists in the very way in which one is inevitably engaged in the world's "ongoing intra-activity" as

Karen Barad (2007) has it (17): "knowing is a matter of part of the world making itself intelligible to another part [. . .] we know because 'we' are *of* the world" (18).

In an interview, Senior offered that she is interested in picturing a non-anthropomorphic universe (deCaires Narain, 1996). Participating in such a world may mean relinquishing the (controlling) self, and a certain surrender (O'Brien, 2002, 176). I do not want to argue that Senior's interest in the premodern and anti-modern would be part of the later project of posthumanism. Social or political ecology would offer a more appropriate frame for reading Senior's nature poetry. Nevertheless, her work does offer a non-anthropomorphic, nonbinary sensitivity that can be highlighted by a modest materialist reading. This reading could be specifically useful to understand the nature of her *writing*. Senior does not evoke the practice of gardening as a metaphor, but an act in its own virtue, that should produce the food that makes it possible to survive. As such, gardening is the very essence of dwelling. So is writing poetry—as situated practice, it stands *next* to gardening, instead of serving as the *signifier* of the trope of gardening.

Here we touch on a central theme in Senior's writing: the insight that practices such as writing and gardening are both strategies of survival, and none has precedence over the other. Language and poetry may help to identify the power relations, but the imagination is less productive than pragmatism when it comes to the unglamorous mundane struggle against the powers that be. Let us face it: beautiful thoughts alone do not necessarily lead to good practice, as many thinkers in this field repeat (e.g., Hallward, 2001, xv, xx; Marshall, 2002, 9–10; O'Brien, 2002, 169). Writing, gardening, and social and political practices are each hazardous practices in their own right. They intervene in the highly specific power-dynamics of their particular field (language, environment, society), and have therefore their own materiality. Writing cannot serve as the universal, transcendent metalanguage that exposes the hidden dynamics in *another* field. Nevertheless, writing has a particular force. As a discursive practice that was shaped by its interaction with the social and natural environment, it is able to question and revise the discursive dimensions of environmental and social activism. Whether and how that discursive intervention results in changes in the practice cannot be predicted.

My point in lingering on the materialism in Senior's writing supports my argument that aspects of a critique of humanism can be found in certain literary traditions that are at the same time engaged in identitarian processes, without any contradiction. Conversely, the critique of representation to be found in new materialism (from Barad to Butler), in favor of agency and performance, need not exclude an interest in identity, or specificity. Senior's understanding of the (human and natural) world as active and intertwining is not related to new materialism or any other antirepresentational philosophy, but to her poetry's political position, which is highly critical of capitalism, and implies her feminism.[24] Thus, if Senior subscribes the materialist notion that the world is intra-active, she will see this less as an epistemological statement than as relating to the specific agency of the natural and the political world.

Moreover, in line with feminist views in the 1980s, Senior understands these political practices as necessarily personal, local, and specific (O'Callaghan, 1988, 147), thus bringing in again the specific element downplayed by some new materialist thinkers, as Hallward (2001) argued. Instead of judging Senior's concern with identity issues as a lack of conceptual rigor, I would like to place her on a map of poetic positions, in the space where the critique of humanism and an approach of the (nonbinary) world as agency go happily together with a focus on the specific—as opposed to those positions (as in the later Glissant) where a critique of humanism and an emphasis on agency (becoming) is explained within what Hallward would call a singularist frame.

Different Caribbean writers, writing in different decades, tie in with the most influential bodies of theory in the study of postcoloniality and globalization nowadays, but offer their own nuances: they prefer not to focus on mobility or expansion alone, or the grand singular forces that shape the totality of the world (as the later Glissant), but stick with the specificity of place, a place approached as a broken landscape, for which one takes one's responsibility nevertheless, even if that means managing the forces of both present-day political and commercial powers, and Deep Time and Outer Space.

NOTES

1. This kind of knowledge that stems from the decision to remain in one place, can be opposed to the Enlightenment concept of knowledge that is connected to travel (Verstraete, 2002).
2. Garrard, 2004, 120–27.
3. A growing number of scholars working in the intersection of ecocriticism and postcolonial theory have offered equally thought-provoking critiques (e.g., Lawrence Buell, Elizabeth DeLoughrey, Graham Huggan, Susie O'Brien). Their criticism often bears on their opponents' inadequate theorizing of the *specificity* of space.
4. My use of the term is not meant to bring in Heidegger's notion of dwelling, as I suggested above. For a short discussion of Heidegger's theory of dwelling, see Garrard, 2004, 110–13; Rigby (2004). I will follow Ingold's use of Heidegger's term.
5. Alison Donnell (2006) also points out that "the locatedness of her (Senior's, IH) work [. . .] enables us to think about the diaspora paradigm in new and interesting ways" (96). Donnell focuses on the way in which Senior's use of, for example, the "maroon figure resists [. . .] interpellation into the agendas of 1990s black diasporic criticism" (101). This resistance concerns especially the "empowerment of self and the expression of agency associated with migration that we find foregrounded from Gilroy to Boyce Davies" (102).
6. An idea explored in Jamaica Kincaid's work, as I showed above.

7. Macnaghten and Urry (1998) associated this perspective with women, as it has been expressed in certain forms of ecofeminism (155). In the twenty-first century, the concept of the Anthropocene was appropriated in literary studies to indicate a broader body of writing. But reflections on scale did not emerge only then. The key-terms of the day in which Senior was writing her poetry were of course widely different from now, but they expressed a similar concern with time and space, though in a less radical way. In the late 1980s and early 1990s, postmodernity, and, more importantly, globalization were characterized by the very seductive notion of time-space compression (David Harvey, 1989; Doreen Massey, 1994). As we have seen, a few years before the concept was first published, in 1985, Senior already offered a wry comment on its implicit assumptions. In the twenty-first century, deep time and cosmic space (and the concept of the Anthropocene) became crucial issues in environmental and ecocritical writing that was inspired by postcolonialism and the study of globalization, and in postcolonial studies. In his famous 2012 essay, Chakrabarty offers the reason why a postcolonial approach alone would no longer be sufficient: "the current conjuncture of globalization and global warming leaves us with the challenge of having to think of human agency over multiple and incommensurable scales at once" (1). According to Chakrabarty, the two earlier figures of the human (the Enlightenment view of man as a citizen entitled to rights; and the "postcolonial-postmodern" figure marked by differences of gender, race, class etc. [2]) no longer suffice to capture the figure that has emerged in this new era: humanity as a collective geological force, that has no ontological dimension, but is mere agency.

8. Denise deCaires Narain (1996) also points at the theme of rural isolation in Senior's work (341).

9. According to Michael Hardt and Antonio Negri (2000), this is the main characteristic of the new Empire of globalization.

10. I understand imperialism here as the general, age old political and economical process of which colonialism is merely a modern form, which emerged in close interaction with capitalism. Globalization can then both be analyzed as the next stage of imperialism that is a successor to colonialism, and as the next stage of capitalism (McLeod, 2000, 8; Loomba, 2005, 9–10).

11. DeLoughrey, Gosson and Handley's (2005) study departs from the awareness, that is so strong in Caribbean writing, that it is only the Caribbean landscape where history can be recovered, as "history . . . has no surviving witness except nature itself" (1).

12. The soil and sea are both the burying place *and* a monument (DeLoughrey et al., 1) or archive of the past.

13. Astrid H. Roemer (1947) was born in Surinam and has lived and worked in both Surinam and the Netherlands. She is a prolific writer, whose first title was published in 1970; her works address issues of emigration, gender, sexuality, and postcoloniality. Her most ambitious work is a trilogy (1996–1998) that could well be considered the first truly postcolonial literary work in the Netherlands (Hoving, 2004). In 2016, Astrid Roemer was awarded the prestigious P. C. Hooftprijs, one of the most important lifetime achievement literary awards in the Netherlands. She is the first Caribbean-Dutch author to receive the award.

14. Which is exactly the reason why Jamaica Kincaid wishes to keep her distance to her family (chapter 4).

15. Mordecai and Wilson, 1990; Edmondson, 1993. Françoise Lionnet (1992) contested this critique.

16. We could also read a reference to a metropolitan form of theorizing that would be too much enchanted with abstract models (though this is not what the poem intends).

17. See Timothy Brennan (2003) for a comparable critique of Hardt and Negri's (2000) *Empire*.

18. Judging the critical interest in Cliff's treatment of land and landscape, Cliff's work lends itself very well for an ecocritical reading (see apart from the discussed critics, also Edlmair, 1999; Hornung and Ruhe, 1998); Lionnet, 1992; Richards 2005; Walters, 1998).

19. One other form of recovery is the more material vodun-derived practice of hanging bottles in a tree, to capture spirits, and thus collect the voices of a polyphonic, lost past (MacDonald-Smythe, 2001, 168).

20. In earlier work, Cliff had already imagined her own passionate gardening as an ambivalent act: it was a battle against her mother and sister, a struggle to gainsay the motherly

accusation that she was cold, infertile, unable to let things grow (Gourdine, 2003; Hoving 2001a).

21. Let me add that Cliff and Senior both wrote much of their work in the 1980s, when revolutionary action was more generally and hopefully embraced as a means to social justice. Their writings are both shaped by their political commitment.

22. I thank Nora Uitterlinden for reminding me of Stuart Hall's wonderfully productive concept again.

23. If the colonial powers were seen as global, the present-day government is also pictured as part of a global exploitative elite (see, for example, "The Knot Garden," 88).

24. And possibly also her religious or spiritual views.

Chapter Seven

Trans-

Nature, Transgression, and Transience in Surinam

Surinamese Dutch poet Cándani writes in a mesmerizing, poetic style on postcolonial loneliness. If her Jamaican colleague Olive Senior evokes the isolation of former colonies in direct, cynical terms, Cándani pictures the pain of isolation in softer hues. Instead of focusing on the relation between personal distress and the political, she addresses the psychological and sensual disorientation of those who emigrated from Surinam as well as those who stayed. This disorientation is primarily sensed as an embodied experience. Thus, Cándani's work could be read as a phenomenology of postcolonial alienation. An inextricable aspect of that phenomenology is the sensually present Surinamese landscape that permeates her poetic narratives to the extent that it shapes its characters and plot. Cándani's characters emerge as vulnerable bodies in need of sensual reconnection not just with other bodies, but also with the natural environment of Surinam that embodies the memory of the transgression, death, and tragic disappearance of other people.

Above, I argued that the postcolonial desire to heal fragmentation risks to erase the diversity of cultural patterns. Concepts of connectedness imply certain notions of sexuality. For example, the dominant evolutionary approach is based on the notion of a heterosexual, reproductive sexuality that creates genealogical relations between all living organisms (see chapters 2 and 5). Models of connectedness based on the notion of the heterosexual family exclude cultural practices (of sexual behavior, partnership, parenting, creating families) that do not follow that pattern. Therefore, these exclusionary models create new forms of fragmentation. In chapter 2, I show how, in Trinidadian-Canadian writer Mootoo's (1996) novel *Cereus Blooms at*

Night, we find a different, nonheteronormative model of connectedness. The colonial strategies of fragmentation along the axis of race and gender are here countered by transgender and transvestite practices of reconnection. If Marjorie Garber (1993) analyzes the transvestite as a figure that marks a social crisis outside the realm of gender and sexuality, *Cereus* shows that the figure of the cross-dresser can also *bridge* gaps and thus overcome that crisis.

In this aspect, *Huis van as* (House of Ashes) (2002), the second novel by Cándani, a writer of Hindustani Surinamese origin, shares a remarkable kinship with *Cereus*.[1, 2] A clear example of the novel's effort to implicate sexual transgression as the *permanent* marker of a structural, shared crisis is found in the following scene. A man intervenes in a row between two women who accuse each other of adultery. Women should not curse at each other, he tells them. "Everyone has a problem in their home, he said. Everyone has a reason for shame" (69). The fact that this intervention is timidly reported by a seven-year-old boy, who is probably the fruit of an incestuous relationship himself, opens the possibility to read this as a reference to a shared transgression of sexual norms, within a postcolonial context. This reading is strengthened when we consider that this line echoes a much-quoted statement in *Cereus.* In *Cereus*, the following passage not only refers to the shared, hidden practice of cross-dressing and to transgender people, but also implicitly to the colonial condition itself. "[Y]ou are not the first or the only one of your kind in this place. You grow up here and you don't realize almost everybody in this place wish they could be somebody or something else?" (258).

This is not the only moment where the two novels show a likeness. The abundance of sensual references to nature is another similarity. So is the way in which the daughter-figure, involved in incest, is placed on the side of paradisiacal nature itself. Another resemblance can be found in the literary devices of doubling and mirroring, so that many characters become each other's mirrors, or fuse. This device is also prominent in Scott's *Aelred's Sin* (chapter 3). Both Mootoo's and Scott's work (especially his earlier novel *Witchbroom* [1992]) have been discussed within the context of transvestism and transgender expression as projects of subversion. Inspired by these two examples from the Anglophone Caribbean, I want to test a comparative reading of Cándani's novel. To what extent can we say that Cándani's novel, written in Dutch, shares a project comparable with these two Anglophone Caribbean novels, which are reputedly much more subversive than Surinamese literary texts? Are all three of these novels involved in a radical deconstruction of alienating binary gender and racial oppositions through the lens of sexual transgression, but opting for different means to do so?: *Cereus* by means of an exploration of the transgressive sexualities produced by or ruled out by colonialism, such as transgender expression, transvestism, and—on another note—incest; *Aelred* by imagining the possibilities of homosexual love within a colonial religious discourse of intimacy and transcendence;

Huis by reconsidering intimacy, as opposed to postcolonial separation and alienation—possibly even including its most extreme form: incest.

What interests me most in these novels is the fact that their literary practices of reconnection are tied in with a specific, radical reimagination of nature. For example, in *Cereus,* as a boundless and transgressive dynamics of hybridizing through endlessly varied contacts between (hermaphrodite and other) plants, insects, and mammals (human beings, especially those fond of wildlife), rather than as a realm of heterosexual procreation. If transvestite and transgender literary characters have the function of overcoming colonial binarisms, it is especially because they problematize the notion of nature that legitimated these restrictive colonial discourses concerning race and gender. How, then, is nature reimagined? Does the process of evoking sexual and gender "transgression" as remedies for social fragmentation also lead to an alternative understanding of nature and the natural?

The chapters in the second part of this book aim to highlight the diverse ways nature is imagined. I take it, then, that the answer to my question will necessarily be specific and change with every literary work, every cultural context. The aim of this chapter is to bring out the cultural diversity of the reimagination of nature through the lens of the literary figure of the cross-dresser, presenting Cándani's work as an example from the Dutch-speaking Caribbean. I will therefore consider the question to what extent this novel's loving evocation of the opaque Surinamese natural environment can be seen as culturally specific. Additionally, to what extent does Cándani's novel fit in with a more general literary tradition? I will also address a rather painful aspect of many other literary appropriations of sexual "transgression," namely whether the figure of the cross-dresser is erased in the effort to accomplish reconnection. Is sexual diversity thus also erased? Does this novel acknowledge the diversity of sexuality, or deny and dismiss it? How can we situate this erasure or acknowledgment, in social or cultural terms? First, I will consider the presence of images of transvestites and other figures who transgress sexual and gender boundaries in Surinamese (migrant) literature. This will allow me to situate Cándani's imagination of the relation between transgressive sexuality and nature within the context of the Surinamese literary response to its social crises.[3]

THE "TRANS" IN DUTCH-LANGUAGE CARIBBEAN WRITING

For Cuban writer Mayra Santos Febres, the figure of an adolescent transvestite struck her as the best representation imaginable of the Caribbean as a whole. The Caribbean *is* transvestism. It can be understood *as*, and *in*, the spectacular strategy of survival of transvestism, letting its opaque, untranslatable body flaunt its tragic desire for an alterity and multiplicity that cannot be expressed in the linearity of dominant colonial discourse. But to what extent does Santos Febres's intriguing description of the region, a description deeply rooted in the flamboyant Hispanophone Caribbean—where, in the late 1920s, subversive writers celebrated cannibalism as an alternative Caribbean epistemology, and where literary experiment and subversion have long blossomed freely—apply to the modest strips of land that were once colonized by the Dutch? At first sight, the answer is sobering. Surinamese (migrant) literature[4] seems hardly interested in the celebration of such transgressions. Even if the best-known Surinamese comedian on Dutch television performs in drag in a weekly segment,[5] the literary cast of transvestite and transgender characters is limited. Nevertheless, some of these literary characters can be understood to represent Dutch-Caribbean experience.

The unhappy protagonist of a 2005 novel by Surinamese Dutch writer Goudzand Nahar feels an affinity with his transvestite colleague Roy Bandman when he reflects on his *own* position as an ethnic outsider. "How could Bandman live with the knowledge that he did not belong anywhere?" (Goudzand Nahar 156, my translation).[6] Poor Roy Bandman. While he is pictured as the tragic outsider par excellence, he is not even granted the small comfort of belonging to the community of intriguing transvestite or transgender (literary or real-life) characters within his own part of the Caribbean. He might have been comforted to hear about Carlos Seinpaal, a well-educated, middle-class, mercurial black person, whose name means "semaphore." Seinpaal is the subject of a short portrait by the white Dutch social scientist and journalist John Jansen van Galen, a frequent visitor to Surinam.[7] Seinpaal, who by the end of the portrait has transitioned to female and lives as a ridiculed recluse in Surinam, sees herself as "a broadcasting station of sorts." She explains: "I use the echo. I receive the echoing response." Sensitive to the many sexist and racist responses to her appearance, she "analyzes society without taking part in it" (576). In stark, short sentences, she announces, "I am not a mixer, I am not a cocktailmaker." In the course of her life, she has defined herself respectively as a Surinamese man (in the Netherlands) who refused to integrate, as black (in the United States), and as Dutch (in Surinam). Just as the literary character Bandman, this sensitive transgender woman expresses her society's general condition. She exposes its way of alienating and excluding people, as in Bandman's case. She exposes its racism and sexism. This is reminiscent of Garber's statement on the literary and cultural function of the image of the transvestite, whose "presence [. . .] marks the trouble spot, indicating the likelihood of a crisis somewhere, elsewhere" (Garber, 1993, 17). Though the trans woman Seinpaal refuses to embrace the hybridity that is often associated with transgender people and transvestites, she does criticize the tenacious racial and sexual identity politics that mark Surinamese society. There are different ways in which we can read her resistance to metaphors of hybridity. It may be an echo of dominant, middle-class Surinamese standards (as the quest for a stable identity, instead of the foregrounding of difference). However, it can also be the insight that the experience of transgender embodiment may be quite *un*ambiguous. Indeed, as Nicole Seymour (2013) observes, transgender expression is not inevitably connected with postmodernism.

Indeed, the standards of the middle class do not organize all reflections about transvestism or transgender issues in Surinam. The anthropologist and cultural theorist Gloria Wekker explains that the Creole gender system in the lower classes of Surinam is much less rigid and hierarchical than that of the West.[8] The African-Surinamese concept of identity is presented as multiple and dynamic (1994, 78), and both men and women would thus have a certain freedom to behave in ways that are traditionally linked to the opposite sex (100). This relative flexibility can be explained through the African American religion Winti, which informs the Surinamese concept of the self. Winti is characterized by "the belief in personified supernatural beings, that can occupy a human being and rule out his consciousness, after which they can reveal past, present and future and heal illnesses of a supernatural character" (Wooding, 1988, 290–91, qtd in Wekker, 1994, 82). People who are occupied by winti may display the behavior that belongs to this god even if they are not in trance. However, people may also have characteristics that are associated with a winti, without ever having been visited by a winti (83–84). Male and female winti may visit both women and men. In addition, the *kra* (soul) and *dyodyo* (protective divine parents) of human beings have both a male and a female aspect (86–89). Thus, women from the working classes might say that a transvestite man has a prominent female winti, such as the mother goddess Aisa (100). Wekker observes that at certain women's parties, one will almost always encounter men who are dressed and behave like women. In addition, the most popular theater plays staged at a certain college would feature transvestite characters. Within this worldview, the transvestite figure would hardly count as radical and subversive.[9]

Most Surinamese writers dealing with the theme of gender transgression, however, adopt middle class values based on fixed, stable gender binaries. We can conclude that there are at least three different discourses in which the transvestite can be framed in Surinamese writing: as the transgressive figure that marks a crisis in society (Garber), as the more radically liberating transgressor who effectually deconstructs binarism (Santos Febres), and as the representation of traditional, indigenous discourses on gender that differ from colonial or nationalist discourse but need not be subversive (Wekker, Strongman). Below, in relation to Cándani's work, I will argue that there are two more discourses through which transgender figures can be understood. It remains to be seen which of these discourses will be mobilized in a particular literary work, and how that will be done.

THROWING OUT THE TRANSVESTITE

To formulate the beginnings of an answer, we can point out that Bandman and Seinpaal share a destiny as indicators of the crises in Surinamese society with a handful of other sexually marginalized characters in Dutch-Caribbean (migrant) literature, all of whom serve to mark a crisis in the postcolonial condition, or migrant experience. As Marjorie Garber (1993) explains:

> The apparently spontaneous or unexpected or supplementary presence of a transvestite figure in a text (whether fiction or history, verbal or visual, imagistic or "real") that does not seem, thematically, to be primarily concerned with gender difference or blurred gender indicates a category crisis elsewhere, an irresolvable conflict or epistemological crux that destabilizes comfortable binarity, and displaces the resulting discomfort onto a figure that already inhabits [. . .] the margin (17).

In Surinamese (migrant) writing this strategy is almost painfully clear. The few transvestites, transgender people, or other sexual border crossers are, after having functioned as the markers and mediators of the plight of the story's straight heroes and heroines, left to their own devices—that is to say, loneliness, exclusion, and death. One might suspect that Surinamese (migrant) writers do not feel any affinity with Mayra Santos Febres's (2004) sharp analysis of the tragic transvestite nature of the Caribbean condition taken as a whole—the impulse of their writing is toward exclusion and erasure of transgender people, and the restoration of balance and wholeness.

The novel in which the transvestite Roy Bandman figures, written by Surinamese-born and Netherlands-based writer Henna Goudzand Nahar, shows that this project is nevertheless ambivalent. The crisis that animates the narrative could be described as the (neo)colonial condition of forced uniformity and the lack of respect for (ethnic) difference. The story explores the tensions between different ethnic and social groups in a remote village on the margins of an unnamed Caribbean nation, easily recognizable as Surinam, by portraying their opinions concerning the military project of the new tar road—Amerindians, Maroons, the military powers (who have taken over the police force), Creoles, the white Catholic priest, Hindustani, Chinese, and the two main protagonists, who both represent a different form of hybridity. The first of these two protagonists is the slight, insecure, indecisive preacher Wijnruit, the portrait of failed masculinity. Wijnruit is the son of a Maroon mother and a Creole father. The second protagonist is the teacher Bandman, a man with feminine beauty. Bandman is a closeted transvestite and used to call himself Samantha, facts known only to his close neighbor and colleague Wijnruit. Bandman, although he feels haunted, excluded, and isolated, is also able to cross borders. He is, for example, able to cross the boundary between the Creole district and the Maroon district with ease, in contrast to his pupils.

The larger historical context of the story is the (ongoing) tension between (especially) the Maroon people and the groups exploiting the regional goldmines. The planned road will lead to greater control of the region (including the goldmines) by the dominant Creole people and the military powers. Conformity was already the norm in a moral sense, "in that small community [. . .] uniformity was made the rule" (101). However, the Creole people and the military forces are now literally forcing the Maroon villagers out in order to take over the exploitation of the region's gold. This brutal form of (neo)colonization is expressed through the transvestite's predicament. If the village is made accessible by the new tar road, the transvestite teacher will also have to flee, for fear of being unmasked by former acquaintances. His transgression, then, is associated with the intermediary spaces between the rural community and its immediate natural environment that have, as yet, not been paved by neocolonial forces. It is further associated with the impenetrable spaces beyond the reach of the neocolonial forces, and with the border crossing between ethnic communities. Transvestism is thus not situated beyond the diverse community's social discourse (it does not belong *in* the jungle, like the lepers in Roemer's novel; see chapter 6) but in its margins, and it is threatened by a brutally invasive neocolonial discourse.

The narrative about unresolvable diversity and opposed interests ends on a note of reconciliation. Through Bandman's tragic example, the preacher understands that individuality is a "lie" and that one needs one's community (156). The Maroon philosophy, which is here conceived of as maternal, is the key to a healthy balance between community and individuality, because it involves a pragmatic tolerance toward diversity, as opposed to the destructive post- or neocolonial strategy. This becomes apparent in the Maroon approach to Bandman's transvestism. When Bandman realizes that his cross-dressing practices have been discovered, he prepares to commit suicide. The sympathetic preacher Wijnruit unexpectedly comes to the rescue to explain Bandman's nightly appearance as a woman in the graveyard just outside his house. Wijnruit invents a story about a dead sister haunting Bandman. The Maroon leader, believing the tale, offers to perform a ritual for Bandman to allow him to lay the sisterly spirit at rest. Afterward, the preacher proposes to let Bandman build a wooden fence around his house, so that he can hide from all the travelers from the capital city that the road will bring to the formerly inaccessible village.

Two discourses concerning transvestism collide. Bandman represents ethnic divisions, which are here imagined as gendered (e.g., through the description of Maroon wisdom as maternal). If Bandman is gently removed from sight, it is because his behavior is seen as incompatible with the desired ethnic and sexual organization of the community. Bandman's gendered complexity is seen as either unspeakable or cursed, and he is urged to dismiss his feminine aspect. In supplementary fashion, Bandman signals a postcolonial crisis—society's ethnic fragmentation, political corruption—without being allowed to suggest a way to deal with it. The novel takes ethnic and gender identities very seriously, unwilling to deconstruct them as construction or performance. Nevertheless, it opens a certain space within colonial discourse where the potentially subversive, (post)colonial act of cross-dressing can intentionally *mask* as indigenous, and thus claim—wrongly, but effectively—a traditional form of respect. But this, of course, is the postcolonial condition. This postmodern cultural practice is misunderstood (as indigenous) and misplaced in a way Spivak would call catachresis (1999, 14; 1993, 59–61, 280–81). Moreover, it is also a Maroon pragmatism: the art of living with several conflicting worldviews at the same time. The novel's most interesting aspect lies perhaps in its decision to leave this tension ironically unresolved. At the same time, however, the narrative solution downplays the potentially subversive nature of the act of cross-dressing.

This literary occurrence of cross-dressing can serve as an example of a more general strategy in Dutch-Surinamese (migrant) novels by women writers.[10] The narrative of ethnic and political divisions is supplemented not just by cross-dressing, but also by other images from the sexual realm. Ethnic and political relations are imagined as gendered, and this translation opens up a productive space for their imaginative exploration. The wish to heal the painfully fragmented community is represented as a (collective) sexual desire and/or the desire to procreate, so that a series of metaphors is called forth, such as pregnancy, fertility, abortion. These metaphors are always connected to family and genealogy. Invariably, there will also be a reflection on productive and unproductive sexualities, including pedophilia. In this respect, these women writers respond to the nationalists who, as Paul Gilroy observed (qtd. in Smyth, 145), tend to define the crisis in black communities as a crisis of black masculinity, which should be solved by restoring the traditional heterosexual family—that is, by reinstating the father as the authoritarian patriarchal head of the family and reducing women to the role of mothers of the nation's sons. Many female Surinamese writers also diagnose the (political) crisis as originating in corrupted gender relations, yet they imagine a very different solution. As masculinity (and certainly aggressively patriarchal masculinity) is seen by these writers as part of the problem, the solution is sought in the evocation of *less* patriarchal ways to explore or heal the rupture—from the imagination of a more vulnerable, or hybrid masculinity (in older men [de Noré, Cándani]; in an ethnically mixed foundling [Roemer]) to the forceful re-creation of new genealogical ties by producing interracial babies, and/or the refusal to procreate (Rahman, Roemer). The focus on family relations situates this shared project within the domestic sphere, which calls for an ethical approach if not a straightforwardly moralist one. Perhaps this is why this writing, often realist, ethical, critical, but also supportive of a normative heterosexual discourse, is less enchanted with deconstruction and subversion, and concerns itself more with diagnosis and healing. The natural environment—which is often imagined as an extension of the home (De Vries, Roemer)—may come in as an ambivalent place of ritual healing, though sometimes the healing takes place through violence and death (the river in De Vries; trees and fire in Roemer; see chapter 6).

In the following section, I will show the extent to which the images of (supposed) sexual corruption and sexual transgression function—or fail to function—as a cultural critique. In the last sections, I will come back to Cándani's novel in order to show that it offers a completely different approach from that proposed by Garber or Santos Febres, but one that is at least as unsettling.

DEAD BABIES AND DYING BORDER-CROSSERS

The narratives surrounding pedophilia show how sexual themes are not just issues to explore in their own right, in connection to parental sexual abuse, but that they are indeed also markers of social crisis. The most extreme heteronormative, patriarchal discourses have tended to misconstruct causal relations between very different non-normative expressions and behaviors, or see them as otherwise interconnected (such as pedophilia and cross-dressing). For this reason, widely divergent themes relating to gender and sexuality can have comparable functions—they are *all* seen as dangerous transgressions. More critical forms of creative writing do offer careful differentiations. Pedophilia does not appear as the sign of deconstructive transgression, in Santos Febres's sense. Well-known short-story writer and novelist Ellen Ombre analyzes pedophilia as the sickening mise-en-abîme of exploitative neocolonial structures, exemplified by child prostitution.[11] Highly talented Surinamese short-story writer Marilyn Simons offers a gripping, sensual evocation of poverty and impotence through the imagination of the sexual abuse of children from the position of the victim, highlighting intense parental neglect. In a story by Surinam-born and Amsterdam-based writer Karin Amatmoekrim (2007) we find an unexpected perspective: that of the old, overly protective mother of a pedophile. The mother is guilty of bringing into the home the child that her son desires, allowing him to satisfy his urges. Finally, at his own request, she kills her son. This narrative perspective on the sexual transgression of the pedophile lays bare the violence involved in the imposition of gender roles and colonial racist religious discourse, since this sexual transgression is pictured in the story as a response to racist, sexist, and religious discipline. An unsympathetic female teacher admonishes the young, artistic boy when he draws a black Jesus. The worried mother, who is his only parent, forcefully keeps him from transgressing the gender norms while he cross-dresses playfully as a child. She also keeps him from crossing the line marking the end of youth and the beginning of adulthood through her overly protective attitude, her advice to quit any jobs he attempts, her insistence on his immaturity, and by facilitating his pedophilic desires. In this way, she forces him to stay with her in an endless oedipal imaginary. In this way, the pedophile's desires, understood as a longing for purity, are presented as a caricature of the Christian command to embrace purity and innocence. His desires also mimic his mother's basically patriarchal, colonial insistence on her son's immaturity. Thus, this sexual transgression can be read as a response to the mother figure's *refusal* to transgress colonial racial and gender boundaries. His monstrosity—he calls himself a monster (135)—reveals the perversion in *the mother's* patriarchal, colonial logic.

The important insight this story has to offer is the observation that colonial patriarchal logic leads to violent interventions in gender and sexual expressions, and, possibly, to violent, exploitative sexual behavior. This is the very point that was also made in chapter 2. The link between sexual transgression and the organization of (post)colonial Surinam finds an echo in longer literary texts by other Surinamese (migrant) writers, where a more sustained exploration is attempted of the way in which transgressive sexuality expresses the unresolvable postcolonial corruption of the nation, or the pain of the transnational family network. Central in these literary responses is, again and again, the concern with the tragic separations imposed by (post)colonialism (especially between "racial" and ethnic groups). One possible way to overcome trauma suggests that the task at hand is to reconnect (Herman). As the racial, ethnic, and class divisions are imagined in terms of gender and sexuality, the reconnection cannot but be sexual. Reconnection therefore very often occurs as sexual transgression. While the (post)colonial separations are often unmasked as perverted in the first place,[12] transgressions of the boundaries are imagined as both healing (since they are intended to bring about reconnection) and perverted (since they are often mimetic in nature, and therefore obey the same logic as the colonial system that created the divisions). Thus, the literary imaginations of transgressions lay bare the perversion in the system, while (defiantly) admitting their own perversity.[13] Even if most of these narratives do not aim to offer a reimagination of nature, they do offer a critical reflection on strategies of naturalization.

Often, the central act of sexual transgression is performed by an outright social outcast, who is hardly tolerated—and in different text, this function may be fulfilled by widely divergent characters as the pedophile, the transvestite, the homosexual. This is very clearly the case in the first novel by the Dutch writer Annette de Vries, whose father is from Surinam, where she grew up. *Scheurbuik* (*Scurvy*, 2002)—which literally translates to *torn belly*[14]—is structured around the themes of sexuality, illness, and death. *Scheurbuik* uses these themes to explore the ambivalences of migration and return. De Vries shows that the most painful secrets and conflicts within the family are connected to politics, in this case the coup d'état of February 25, 1980. But De Vries's family narrative supports a more central story line, which deals with the reconciliation between a Dutch-Surinamese woman and the country of her birth. This reconciliation takes place mainly through the reconnection with an old friend, who is dying of leukemia. This friend is the story's most explicit border-crosser. The friend, a gay dancer balancing between life and death, Surinam and Holland, advises the heroine to create a syncretic diasporic (or rather: transnational) identity by bringing together separate elements of Surinamese and Dutch culture, that is to say, by furnishing her house with disparate furniture. When the heroine objects that she does not live in a house but on a ship, and that the earth beneath her feet is shifting, he advises: "Furnish your ship, then. And see that you get your sea legs" (De Vries, 310, my translation). In this friend, we find the paradigmatic border-crosser: a dancer between life and death, masculinity and femininity, ex-colonized and ex-colonizer. This figure is an advocate of syncretism. The dancer clearly points the way for all migrants and children born of interracial couples, who try to negotiate such boundaries. His advice connects Dutch symbols of national self-identification to Caribbean ones, which is especially relevant in this migrant novel. If the *grande dame* of Surinamese writing, Astrid H. Roemer, bases her trilogy on the imperative to conquer one's disgust of leprosy and befriend lepers as the means to face the corruption of postcolonial Surinam (see chapter 6), De Vries offers another fatal physical illness as the symbolic means through which the ever-shifting condition of migration can be accepted. Not just transgressive sexuality, but also the diseased, corrupted body can mark the postcolonial crisis, as well as the means to overcome it. De Vries's ethics are clear: reconnection can only occur when corruption, transgression, and illness are embraced. This story quite explicitly insists that both heterosexuality and homosexuality are binding forces in the transnational Surinamese network of family and friends. Thus, De Vries revises the traditional and nationalist concept of the Surinamese nation as a multigenerational family.

At first sight, the strategy of embracing the transgressive as a means to reconnect seems successful. Remarkably, it is nevertheless the lonely and wise gay border-crosser who has to die to allow the heterosexual heroine to live her full, reconciled life. This makes us pause. Even if the cultural critique produced in these imaginations of sexual transgression is profound, the fact that the transgressive forces themselves (the transvestite, the pedophile, the gay dancer) are eliminated suggests that ethical project of reconnecting presented in these novels frustrates their critical project of questioning society's binary social structure.

The explicit solutions these stories offer to heal society's fragmentation vary from the plea for a syncretic identity that embraces all aspects of one's identity, to the more or less benevolent marginalization (and pathologization) of non-normative behavior. The syncretism propagated by the dying gay dancer in De Vries's novel seems to refer to well-known models of creolization and assimilation (Alba and Nee, 2003) geared toward integration in society, and not at the subversion of the binarism of social norms. De Vries's novel is in the first place about *how* to build a house—or ship—or identity. It is not about questioning the notion of identity itself. But perhaps this is not the most urgent social task for Surinamese writers at this point in time. One might argue that Surinamese writing has not yet exhausted the historical project of re-creating Surinamese history. Dominant Dutch historiography has only recently begun to truly address Dutch colonial history and slavery, and the relation between the Netherlands and Surinam (and the rest of the Dutch Caribbean) is still a highly contested topic. In this context, there is predominantly a great need for stories to address that infamous history—this includes stories that aim to *create* rather than deconstruct an identity.

One can nevertheless discern a cultural critique by taking into account aspects of the narrative which exceed the explicit aim of these stories—that is to say, the project ultimately concerned with the construction of a stable and normative identity. Now the scene is set to bring out the specific position taken by Cándani's novel. Here the tensions around the issue of identity take on an even more complicated pattern, to the point where identity and binarism no longer seem relevant issues. I will also relate Cándani's critique to her depiction of the natural landscape.

READING FOR THE "TRANS-" 1: TRACING AN OEDIPAL TRANSGRESSION

The stories I have discussed so far are organized according to the logic of abjection—the transgressive markers of crisis have to be gotten rid of once they have served their purpose.[15] In some novels, border-crossing characters signify a more permanent alienation that is shared by all citizens of the postcolonial nation, suggesting that the crisis to be faced is much more than just a personal rite of passage. The latter is the case in Hindustani Dutch writer Cándani's second novel *Huis van as* (House of Ashes) (2002).

None of the characters in *Huis* is transgender or a transvestite. Is it possible, then, to suggest that, even if *Huis* does not present these figures, it still offers characters and moments that can be called transgressive, that question and subvert social binarism every bit as much as other novels in which transvestism or transgender experiences are the central theme? Let us try to test this thesis and see where such a reading strategy (a reading for the *trans-* of subversive transgression, even in a novel that does not feature transvestites and transgender people) leads us in the case of Cándani's poetic, evocative novel. A novel in which the Surinamese environment plays a decisive role. It is undeniable that the sexual transgression in this story is severe. The novel's naive narrator, a Hindustani Surinamese girl who returns to Surinam for a visit after spending seven years in the Netherlands, discovers that her own father fathered a son with her bosom friend Señorita when Señorita was only thirteen years old. Furthermore, it is quite possible that Señorita was her father's illegitimate daughter, which would make him both the father and the grandfather of his illegitimate son. The novel relates the daughter's reluctant inquiry into this hidden family history. Her search becomes entwined with her growing despair about her position as an outsider—to her own family, to Surinam, to life itself.

As in other Caribbean novels dealing with these themes, the narrative links transgressive sexual behavior to postcolonial poverty, lack of education, and alcoholism. In this case, however, we are invited to think of incest as a *solution* to these disasters. By picturing the events through the eyes of the returning daughter, the novel offers an unexpected perspective on the oppressive incestuous relationship. Seen from the daughter's point of view, the narrative suggests that it is the twelve-year-old Señorita who seduces a man her father's age, in order to escape the violence of her own alcoholic and abusive biological father. The readers are assured that this relationship turns into a genuine, loving relationship that lasts until the narrator's father's death. It cannot even be defined as incestuous, since the father is not aware of the fact that the girl he loves might be his own daughter until the very last moment. The crisis addressed is thus not primarily the individual male sexual violence in Caribbean society. However, patriarchal social structures *are* criticized, and presented as the cause of the father's transgression as well as the transgression committed by the girl he loves. Patriarchal social structures are also presented as the cause of the narrator's problems. The narrator's alienation from life itself is the result of her mindless decision to let her overly protective father guide her, as patriarchal structures seemed to demand ("with every step I took, I belonged more and more to you [the father]. Less and less I belonged to myself" [94]). Señorita's decision to seduce her friend's father sprang from her wish to live a life other than her mother's, escaping the female destiny of being bullied by parents-in-law, trapped in a loveless marriage. Señorita even states that she wanted to be loved so that she would be able to love herself. More importantly, the novel is organized around the tragic consequences of decolonization. The nation is split by the departure of many citizens, who are accused of abandoning and forgetting those who stay behind (62). Since the novel understands exclusion and isolation as the most destructive conditions imaginable, independence is depicted as a cruel machine that drives some into isolation by fleeing overseas. At the same time, others are condemned to isolation because they are left behind. Against expectations, the novel does not consider the father's transgressive sexual behavior as the novel's main transgression. The main transgression, rather, lies in the father's decision to migrate to the Netherlands to escape the effects of his deeds, and to create a safe distance between his innocent daughter and the disgrace of his sexual behavior in Surinam, which becomes evident through Señorita's pregnancy. Thus he has forced both girls into a deathly isolation. Tellingly, the father legitimizes his (sexually motivated) departure by his dissatisfaction with independence. His flight from Surinam is the migrant's anxious effort to distance himself from the suffocating and corrupt sexual politics and gender politics in the former colony. This includes his own scandalous behavior. This effort echoes Césaire's evocation of the Caribbean island as the corrupted maternal body.

An allegorical reading of an incident in the father's youth illustrates this point. When young, the narrator's father is lost for three days in the depths of the forest during a hunting party. When he returns, he has turned wild and no longer speaks. It is said that his real self stayed in the forest, and what had returned was a body inhabited by another (98). The father begins to travel. He returns to Surinam and then departs to the Netherlands, where he finally dies in a car accident. The corrupted father can be said to embody Surinam itself when he was young, and the alienation from Surinam when older. The fact that the alienated and heartbroken father recognizes himself in the simple Señorita, who is represented as life itself, supports this reading. Señorita is that very authenticity he has lost during his mysterious three-day stay in the tropical forest. Señorita is closely linked to the half-cultivated, sensual Surinamese landscape of the meadows next to the house. She pops up, mysteriously, at the very beginning of the story, when the prodigal daughter contemplates the way in which nature has reclaimed the cultivated fields around her former home. She recognizes the sprawling plants she had once planted herself, which now grow jumbled and wild. Butterflies welcome her, smells surround her, the wind opens the door of the old home, until she feels she is being watched. From the heart of this friendly, forgotten, exuberant nature, Señorita appears like a mirage, a mirror image, as nature's soul itself—or the soul the narrator has lost (7–9). It is Señorita who tells the narrator the names of the flowers she has forgotten (31). Later, when contemplating the happy lushness of nature, the misled daughter sighs: "—all this is now assaulted. By loving Señorita you have assaulted everything. My youth. My memories" (57). Through this statement, Señorita all but vanishes in the narrator's projection of her own happy youth, symbolized by this natural paradise, on Señorita's person.

Leaving the connection between Señorita and the lush landscape aside for the moment, I will elaborate on the oedipal story that seems to dominate the narrative. This narrative is clearly marked by the narrator's oedipal desire for the father. Reconnecting with her father—and through him also with Surinam—would offer the narrator a full subject position, marked by gender, sexuality, and nationality. Remarkably, the novel pictures the *migrant*'s life in the privileged zones of Western Europe as sterile—instead of that of the impoverished inhabitants of the marginal spaces in Surinam, who struggle with poverty, alcoholism, and broken families. To the narrator, life and fullness are directly connected to the non-normative love, childbirth, and humiliation one finds in the deprived areas of Surinam—and in the figure of Señorita (87). Feeling excluded and disregarding social norms, the returning daughter merely longs to be marked by life, love, and suffering. She desperately wants to "[b]ecome one with everything, in the way I could be one with Señorita in the old times" (27). As she realizes only now, she never knew the father she followed and loved blindly. She now learns that she cannot relate to her father if she does not acknowledge his hidden sexual history. She cannot embrace Surinam if she does not embrace its hidden histories of incest. Señorita functions as the mirror that shows her that incest is at the heart of her desire for the father—and this, in my eyes, is the transgressive moment of the narrative. All the more so, because Señorita refuses to obey the prohibition. The narrator wavers between a narrative according to the logic of the imaginary. Her intense need for reconnection makes her see the world as organized around the binary opposition of closeness/isolation only, so that adultery and incest become meaningless words—and a narrative according to the logic of the symbolic, where adultery and incest *do* count as transgressions she must condemn. Without the alienation caused by these taboos, one cannot enter the symbolic order. But the narrator, sick with the desire of reunion, has trouble accepting this.

On the one hand, then, we discern a radical resistance against the (neo)colonial patriarchal family values that destroy the (in this novel) essential value of intimacy, which is tied to the imaginary. In the desire for closeness, there is no such thing as transgressive love, for there are no boundaries to transgress. There is only the sterility of distance, contrasted with the fullness of intimacy and connectedness. On the other hand, the novel obeys a heterosexual, oedipal logic. This second pattern checks the story's subversiveness, and forms an obstacle to the emergence of transgender characters, transvestites, or lesbian love. The crisis addressed by the incestuous girl Señorita is that of postcolonial isolation. She embodies both the bliss of connectedness, and the horrifying transgression of incest that is nothing but the extreme consequence of the desire for connectedness, which originates in the traumatic divisions caused by colonialism.

READING FOR THE "TRANS-" 2: IN THE VERY HEART OF BINARISM

Earlier, I suggested that this Dutch-Surinamese migrant novel is less subversive than its Anglophone counterparts, since, by refraining from portraying queer and/or transgender characters, it does not explode dominant heterosexual, oedipal patterns. However, let us try out an experimental reading, and ask whether it is not possible to discern the transvestite figure in Señorita? The girl's seductive behavior is in part modeled after an adulterous woman she once spied on. It is even more explicitly modeled after the movie stars and fictional heroines she imitates when having sex. "I played how they would play. Looking drunkenly and deeply into the man's eyes. I talked with fingers, because I did not have the beautiful words" (43).[16] This young FTF-transvestite[17] understands that femininity is a role to play. Without words, she grasps the essence of the notion of simulacrum, through which, according to Severo Sarduy (1989), we can understand transvestism (97). As Sarduy argues, with reference to Baudrillard's concept, simulacra work without referring to the original Platonic Idea, or an inner essence. For Marjorie Garber (1993), this is the reason to suggest that "the transvestite marks the entry into the Symbolic" (354). Indeed, not as a transvestite presenting femininity, but as a transvestite who reflects on her performance, Señorita clearly understands that femininity and masculinity are nothing but performances, social codes. Señorita's acceptance of the coded nature of the symbolic order makes her much more part of that order than the narrator, who is the real outsider.

However, close scrutiny suggests that play on performativity is not the novel's main concern. Rather, there is an imaginary logic at work in Cándani's narrative, especially because it is presented through the eyes of the focalizing narrator, who longs for seamless reconnection. The story displays an infinite mirroring and doubling of all characters, without ever referring to a stabilizing, foundational power, in the same way that *Cereus* and *Aelred* also do not do. Many characters seem to melt into each other, functioning as each other's mirrors or shadows. When Cándani's narrator walks toward her bosom friend Señorita for the first time after seven years, it is "as if I was walking towards myself" (9). She sees her as a clay puppet, made "to create the woman I had loved to be myself" (13), and remembers how she learned to imitate her friend's voice to the point where even Señorita's mother was fooled. On the other hand, she also complains that "[Señorita's] full life mirrored my own empty life" (45–46). Sometimes the two girls merge into a single figure, or shared body ("the girl with whom I formed one whole" [9], "She lies so close as if her breathing passes through my lungs" [77]). At other times, they are each other's mirror image. One girl represents the uneducated, poor, sexually transgressive, but authentic and full life of postcolonial Surinam, the other the educated, rich, sexually experienced but unmarked, alienated, empty life of the migrated Surinamese. The game of mirroring does not stop there. Sometimes the narrator seems to take on her father's position. Sometimes she even positions herself as Señorita's lover, as if claiming the son's oedipal right to love the woman his father loves ("My heart desires her skin. [. . .] Father, my father has loved the magnificent skin of b-e-a-u-t-i-f-u-l-l-a-d-y" [35–36]).

The parallels and mirror images keep proliferating. To Señorita's grandmother, both her own child (Señorita's father) and the narrator's father are own sons. She has breastfed both (101). Does that mean that Señorita's desire for her friend's father is fundamentally an incestuous desire, because her friend's father can already be considered as equal to a father to her, long before it is revealed that she might also be his biological daughter? And do all women collapse into one? Indeed, the narrator sees Señorita and herself in Señorita's grandmother. Señorita's grandmother in turn mirrors all women, as many women in Surinam will become grandmothers one day (84). What should we make of the fact that Señorita's name, given to her by mistake by her inadequate father, is an empty name? And what should we make of the fact that the narrator deconstructs that name (see the play with letters on page 35–36 quoted above)? Her son Anjana's name means "the unknown," but it is also his father (who is also the father of one or possibly both young women) who is described as "the unknown." A last important example of mirroring can be found in Señorita's identification with her great-aunt, who committed suicide when she found herself pregnant as the result of an incestuous act. When Señorita was born, her grandmother (the deceased's sister) said "that her sister had returned" (38).

The hut where Señorita's great-aunt commits suicide, and where all significant and transgressive actions of the narrative take place—this includes its natural surroundings, where meaningful patrilineal names are lacking—can be read in Lacanian fashion as the timeless imaginary space where the protagonists are held captive within an endless mirroring. We might understand this imagination as the utopian space where transgressive (incestuous) desires can be satisfied, where girls can immerse themselves in amorous relationships that remain outside the logic of patriarchy. The imaginary is, as Kaja Silverman (1983) explains the Lacanian term, "that order of the subject's experience which is dominated by identification and duality. . . . [It makes it] possible to discover correspondences and homologies" (157). In contrast to the symbolic order, it is the place where one is not (yet) fully alienated from one's own needs (Silverman, 162). Following Silverman's and Spivak's insistence that the imaginary is not merely a stage to be resolved and left behind, but a register that coexists with the symbolic order throughout life, we might recognize this narrative space as a Caribbean space of utopian transgression—one where feminine and masculine roles are diminished—the prepubescent girl is not yet a woman when she seduces the man, the man has renounced his active masculinity. This is reminiscent of the lovers in *Cereus*, a transgender boy who is no longer a girl, and a cross-dresser who has renounced his masculinity. However, in Cándani´s novel the gender roles are not subverted or reversed. Safely situated in a space outside the realm of the Law, in the very heart of binarism and duality, we discover a utopian play with binary gender patterns. But this play is fulfilling within that space only, as the players turn out to be victims or losers as soon as they (re?)emerge in the symbolic order.

There are two reasons, then, why we could question the social subversiveness of this transvestite play with binary codes. The first reason is that my effort to read Señorita as a transvestite shows that the novel reveals only *some* of the insights embodied by the transvestite (for example, the insight that gender roles are performative), and not the radical insight into the very real violence of a rigid gender system that comes with the literal transgressions of gender boundaries. The second reason why the transvestite would not be subversive lies in the more general observation that the play with binary-crossing offers no solution to any social or political (post)colonial crisis.

An anthology on postcolonial and queer theory quotes Hardt and Negri: "the postmodernist and postcolonialist [and queer?] theorists who advocate a politics of difference, fluidity, and hybridity in order to challenge the binaries and essentialism of modern sovereignty have been outflanked by the strategies of power" (138, qtd. in Hawley, 2001, 9, his comment). The transvestite and/or the border-crosser often functions as the very embodiment of the concept of hybridity. However, that concept has been exhaustingly discussed as possibly counterproductive, as it is tenaciously steeped in a dialectics that leads to synthesis, instead of taking its distance from, or questioning, binarism (Werbner and Modood, 1997; Purdom, 1995).[18] Yet, as many Caribbean theorists such as Edouard Glissant have argued, the divisions of the postcolonial condition can not be healed. The history of colonialism and slavery has ruptured the continuity of genealogy that supported the community. As a result, the Caribbean has become an area of incommensurable, untranslatable differences that cannot be known according to the logic of legitimacy. The opacity of the postcolonial world is not *knowable*. It is precisely this insight, also presented in Cándani's novel, that opens the possibility for a new way of reading. This reading allows us to recognize a different narrative in Cándani's novel, and to understand what cultural work the novel's transgressive character is accomplishing. Indeed, the subversion of social categories is not its main effect. But it performs another kind of cultural critique. In the section below, I will offer a nonpsychoanalytical, nonbinary reading, inspired by posthumanism, ecocriticism, and insights from transgender studies, and move away from the tenacious binarism of the (post)colonial imaginary.

READING FOR THE "TRANS-" 3: BEYOND HYBRIDITY

When Santos Febres looks at Glissant for an alternative imagination, that does not obey the logic of legitimacy, she finds it in his poetics of Relation, which closely resembles the logic of illegitimacy represented by her own notion of the transvestite. Santos Febres takes care to point out that Glissant's relationality represents "a synthesis that can not be reached through reunion, but through expansion." With this term, a sudden link is established between Cándani's narrative and Deleuzian-Glissantian theory.

My psychoanalytical reading above does not take into account the prominence of nature in the narrative. Much more than a mere setting, it is an agent in its own right. The overwhelming natural sensuality that forms the backdrop of the story is another aspect the narrative has in common with *Cereus* and *Aelred*. Above, I proposed a reading of this celebration of nature as an allegorical story of the migrant's longing for her *reunion* with the lost nation. Here, I will approach it as a narrative of joyful *expansion*, in the Glissantian sense, where a young girl realizes her desires by entering a multiplicity of relationships, with the human and nonhuman alike. Rather than understanding the endless game of mirroring and doubling as a play within the imaginary, I would now like to propose to see the game as an effort to *pluralize* the body. To a certain extent, this desire for multiplicity is already implicit in Santos Febres's definition of the transvestite, as quoted at the start of this essay. However, in feminist new materialist approaches this plurality receives much more emphasis. Moira Gatens (2000), for example, defines the "body as a nexus of variable interconnections, a multiplicity" (61). The novel allows us to differentiate two alternative views on Señorita's erotic behavior. The narrator, still taking the paternal Law seriously, is shocked to see her sensual friend dismiss it. The friend, not very interested in the laws, since she has already accepted that all laws are mere codes prescribing performance, is seduced by the possibilities of "becoming" in itself. To stay with Deleuzian terms, Señorita can be seen to inhabit the plane of immanence, the plane of becoming, while the narrator is situated on the plane of transcendence. For the narrator, desire obeys a Lacanian psychoanalytical logic. It is, as Elizabeth Grosz (1993) explains, "that yearning to fill in, to reproduce a lost plenitude" (171). At the same time, to Señorita, desire seems to be something else: "[I]nstead of a yearning, desire is an actualization, a series of practices, action, production, bringing together components, making machines, making reality" (ibid.). Now we can understand why Señorita, in spite of her own worries about her isolation and lack of education, seems both a more passionate and a more pragmatical personality than the nostalgic, desperately longing narrator. In Grosz's rendering of the Deleuzian view, "[d]esire does not take for itself a particular object whose attainment it requires; rather, it aims at nothing in particular above and beyond its own proliferation or self-expansion" (172). One remembers that Señorita desired the father to be able to love *herself*. Her desire is concerned with *loving*—it did not emerge as the desire for a particular love object.

This reading becomes manifest when we respect the novel's careful evocation of Señorita as a sensual body. Señorita is a beautiful, sensual appearance, a tender body that haunts the narrator. The narrator loves to sleep in the same bed with Señorita. The girl is a living part of the sparkling, scented landscape that surrounds her house, which is also significant and deeply moving to the narrator. The narrator observes the landscape continuously, and with the same intensity with which she observes her friend. In feminist appropriations of Deleuzian theory, we find a radical redefinition of subjectivity through a return to its material basis: the human body itself.[19] As Claire Colebrook explains, "[t]he body *is* in its modes of practice, self-representation, and engagement" (2000, 86). The body should be seen in the Deleuzian terms of *becoming,* so that the human can be seen as "*nothing other than* an interpretation of its own body (a becoming-other than the body)" (86). In Cándani's novel, this emphasis on the corporeal becoming of human subjectivity helps us to see that Señorita is not just imagined as a girl in lush surroundings. She is represented as a sensual organism within a sensual network of plant and animal life, which is imagined as the place of becoming itself. It is here that the narrator becomes aware, again and again, of the fact that the girl is not her projection, her dream twin, her creation, but "a woman" in her own right—a description I wish to read as a synonym for a sexual, desiring subject. Explaining Moira Gatens's political, feminist reading of the Deleuzian-Spinozist notion of subjectivity, Colebrook states that Gatens sees "the sexual subject as an effect of *doubling*, whereby the subject occurs as a relation to its image. [. . .] It is the doubling of the material body *as* an ideal body. It is the material becoming other than itself; and this becoming also occurs in relation to an other body" (Colebrook 86). Indeed, Señorita seduces the narrator's father by imitating the pictures of beautiful women she has seen. If we explain this imitation as an instance of doubling, which is then repeated in the very many other instances of doubling (Señorita as the incestuous great-aunt who killed herself, Señorita as the narrator, Señorita as the Surinamese landscape, and so on), an image emerges of a girl who is constantly becoming in relation to other bodies.

My point is that the significance of this narrative does not lie in demonstrating that gender identities are merely performative. There is still reason to see Señorita as a transvestite. Not because she presents a visual spectacle, but because she adopts the small gestures of seduction we can also find in the landscape surrounding her: the butterfly touch. Here we see the affinity not just with *Cereus*'s "insect aesthetics," but also with the approach to transvestism in Sarduy, which is based on studies of animal mimesis. When the writer and a friend meet a group of transvestites who are like "phosphorescent bands of recently opened chrysalises," he is "brushed, as if by lethal elytra, by the sharp, starched organzas of pleated skirts, the poisoned flowers clutched in skinny yellowish hands, and also the cheap, sickeningly sweet perfume" (94). This is not a transvestism of the gaze, but a transvestism of the touch and smell. It is not meant to subvert, but to relate, seduce, and, in Señorita's case, heal separation and loneliness. In the touch, we find a discourse of healing that is very closely related to Glissant's poetics of Relation. It offers an intimate way of knowing that is not intellectual, but that, in Cándani's work, represents another healing model of postcolonial or migrant being. In her first novel, *Oude onbekenden* (Old unknown people, or Old unacquaintances), the young female protagonist, a migrant, realizes a desired return by entering into amorous relationships with Surinamese men. She is prompted by a desire to relate to their bodies. Her relationships represent an effort to pluralize herself, to relate, to expand.[20]

It is this emphasis on touch and smell that *Huis van as* brings to light, an insight that is also scrutinized in transgender studies. Susan Stryker explains that transgender studies is "more attuned to questions of embodiment and identity than to those of desire and sexuality" (as queer studies would be) (2006, 7). The interest in the materiality of human existence also defines this literary imagination. The novel opens a space where the body can be lovingly explored as unknown and unknowable—a body exceeding cultural systems of signification, and that is resistant to many forms of categorizing and essentializing. Even while Señorita is firmly categorized as a woman (which complicates a sustained transgender reading), she exceeds the legal categories of daughter/sister/mother/victim. At the end of the novel, however, Señorita willingly seeks to bring the sensual relation between her body and her living surroundings under what Stryker calls, following Foucault, a "regime of normalization" (Stryker, 13). It is the narrator's gaze that insists on the opaque complexity of her friend's embodied existence.

To recapitulate, the narrative offers a subtle interweaving of different narratives of sexual transgression, which work with and against each other. Firstly, we have the identitarian oedipal account of the narrator, who judges the events under the denominator of perverted transgression, emphasizing the decay that is the result of the transgression (an evaluation that is expressed in the title, *House of Ashes*), but who is nevertheless jealous of the transgressors. In this narrative, the incestuous transgression emerges as the sign of a category-crisis (between generations, between [neo]colonial classes, between nationalities) in a way that corresponds strongly with the significance Garber attributes to the transvestite figure. It threatens the narrator's own conventional existence. Secondly, and closely related, the narrator evokes an imaginary logic that she projects on Señorita, and creates a space for her own dreams of reunion. Thirdly, there is the narrative of desire suggested by

Señorita, which is not based on oedipal norms, but on an ethics of becoming.[21] Señorita's narrative might be understood as a continuation of Santos Febres's notion of the transvestite, as Señorita is marked by the transvestite's desire of multiplicity and alterity, but refuses (in the Deleuzian-Glissantian sense) the tragedy in this desire. This refusal becomes visible in her dismissal of the division between the imaginary and the symbolic, an attitude expressed by her pragmatic desire of a "normal" social life, schooling, and the like, without renouncing her distance to patriarchal norms. Señorita's narrative is not organized around an identity in crisis, as in Judith Butler's work, but is concerned with the affirmative nature of becoming (Colebrook, 2000, 90; Grosz, 1993). As I argued above, the emphasis on embodiment and materiality in this last new materialist approach ties in with a crucial insight from transgender studies.

But this is not all. In the fourth place, we can discern a place-bound, spiritual discourse—firstly, expressed through Señorita's grandmother's mild tolerance, based on her belief that Señorita is her returned sister; secondly, in the representation of nature itself. Within the story, nature functions as an alternative discourse, with its own ethics. The narrator visits her father's grave: "Silent and quiet he lies lying, and smells of the earth that has received him. More peaceful than when he would lie in God's arms, where he would have to give account. A scent, a breath he is now, together with all the others here, the plants, the wind, in the sun and the rain" (106). One recognizes in this quote—and in other references in the novel to a desire to escape social constraints and to become one with life itself (27, 78, 103)—the generalized ideal of deliverance from the confinements of human social life (which occurs in both Hinduism and Buddhism).[22] The novel translates this ideal into a valorization of nature as the vital space where life unfolds freely, unhampered by loveless prohibitions or binary logic. The image of the butterfly embodies this freedom, which could be called transgressive if it were not based on the dismissal of boundaries as futile, within a natural discourse. On the one hand, we might venture to see this discourse, which celebrates nature's boundless plurality and mobility, as related to the Hindustani critique of creole dominance in the definition of the decolonizing nation of Surinam since the 1950s. Hindustani critics emphasized the cultural *diversity* of the nation, adopting the Dutch model of compartmentalization, striving for Hindustani institutions (Choenni and Adhin, 2003, 107). On the other hand, the novel's plea for plurality is also pitted *against* the Hindustani ethnic institutions, especially patriarchal marriage. Diversity and hybridity in themselves do not necessarily lead to freedom. Nor do images of hybridity, such as the literary imagination of the transvestite, necessarily signify subversion. All forms of diversity and hybridity may turn into confining institutions in their turn. At the very end of the story, the narrator explicitly embraces a nonbinary ethics of radical multiplicity, relationality, and expanding potentiality, that is associated with mobility. Señorita's little son sends her a drawing of his new cat, and tells her about his new friends: "A new player. Time goes by, brings with it new lives" (108). Movement, not the frozen framework of oedipal logic, is the vitalizing principle. When she looks at the whirling leaves, she muses: "Even the dead, I think, do not persist in lying still" (108).[23]

Neither do the many images of sexual transgression, in Dutch-Caribbean writing. The different, conflicting images keep marking the moments of pain and violence in the postcolonial and migrant experience, sometimes questioning binarism, sometimes reinforcing it, while they check and challenge each other. Different, sometimes incommensurable, postmodern, Caribbean postcolonial, new materialist, nationalist, and traditional Surinamese or other spiritual or religious notions are evoked in an ongoing literary reflection that tries to relate these discourses to each other in a *syncretic* way,[24] respecting their irreconcilable differences. The literary explorations of these varied discourses serve different social aims, including historiography, healing, cultural critique, social comment, ethical innovation, and sometimes subversion. However, as these discourses are not always organized around binaries of race, gender, and class, the transvestite is not necessarily subversive. S/he is also not the only image of sexual transgression to perform a cultural critique. What seems to emerge as a shared characteristic of this specific body of writing, is the fact that these Surinamese (migrant) writers diagnose the social crisis not so much as an effect of binarism, but rather as an effect of isolation, divisions, failed relations, and responsibility. The task at hand, then, is the search for new ways to reconnect—or new ways to relate to the irreparable dividedness of postcoloniality, in a Glissantian and/or posthumanist, materialist way, or through an appeal to non-Western, nonhumanist epistemologies of being. The imagination of nature does not serve as a radical, political critique of notions of the natural. Instead, the natural landscape leads to an ethical insight in the transience of life, and the existential anguish of alienated, fragmented lives. Curiously, even though this insight may sound like a universalist truism, the observation appears to be shaped by a *specific* (partly Hindustani) imagination of a *specific* condition—a condition that is marked by postcolonial, rural poverty and the fragmentation caused by recent emigration from a nation in the process of creating its postcolonial identity.

NOTES

1. Poet and novelist Cándani (1965) was born in Surinam, and migrated to the Netherlands in 1990. She writes both in Sarnami and Dutch.
2. The Hindustani came to Surinam in the last three decades of the nineteenth century, as contract laborers from North India, then a British colony. They now represent more than a quarter of the Surinamese population.

3. It may be useful to add that this chapter started out as a broader exploration of the role of cross-dressing in Dutch-Caribbean literature. However, I managed to find only a small number of characters that could count as cross-dressers, and still fewer helped shape a cultural critique of colonial, patriarchal patterns. I then decided to analyze a broader range of Dutch-Caribbean imaginations of non-normative expressions of gender and sexuality, and their relation to notions of nature and the natural. As the reader of this chapter will find, many of these imaginations are marked by a still active colonial heritage; they testify to deep-seated social anxieties, and often the literary characters embodying non-normativity are sacrificed to solve that anxiety. For an analysis of more critical Caribbean imaginations of the cross-dresser see Fumagalli, Ledent, and del Valle Alcalá (2013); their project inspired the writing of this chapter, and I thank Benedicte Ledent and her colleagues for suggesting this rich theme to me.

4. In this chapter, I refer to novels written by writers who were born in Surinam and still live there, novels by writers who were born in Surinam but have migrated to the Netherlands, and to novels by Dutch immigrant writers from a Surinamese family background. Though I depart from the assumption that, in spite of their many differences, there is a continuity between the works of these writers, I will try to be specific about each writer's background.

5. The Surinamese-Dutch comedian Jörgen Raymann has his own weekly television show, in which "Auntie Es," an elderly Surinamese woman, is one of his most popular impersonations.

6. All translations of Dutch texts are mine.

7. I am grateful to Michiel van Kempen for guiding me to this fascinating short text. It appeared in a collection of short portraits, essays, and anecdotes that portrayed the state of decolonizing Surinam and the difficult Dutch-Surinamese relations (John Jansen van Galen, *Kapotte plantage*). Within that publication, it consisted of several brief fragments that were interwoven with other essays, suggesting that the fragmented life narrative of this transvestite is an intrinsic part of a fragmented, multiple Surinamese history. The story was later published as a coherent whole in Michiel van Kempen and Wim Rutgers's (2005) *Noordoostpassanten*.

8. See also Kamala Kempadoo (2008), who points at the perhaps well-known argument that, in the Caribbean, sexuality is less seen as an identity than a practice; a man who has sex with another man will not necessarily identify as gay because of that practice (28–29), though this depends on the specific context.

9. See also Gloria Wekker, 1994 and 2006; Strongman (2007) for the relation between conceptions of gender and sexuality and Afro-Diasporic religions, notably Vodou; and Alexander (2006), for a probing analysis of alternative histories of women's sexuality, and the role of spirituality.

10. Let me add that the overwhelming majority of the Surinamese or Surinamese-Dutch writers who, at the start of the twenty-first century, publish in Dutch is female.

11. Ellen Ombre (1948) was born in Surinam, came to the Netherlands in 1961, and returned to Surinam a few years ago.

12. Robert Young (1995) has theorized the white obsession with illegitimate sexuality in racial theories.

13. Though in that case, the marginalized literary character may well be sacrificed to protect the existing, dominant system.

14. In Surinamese words, this expression also refers to the condition of the *bere*, the "belly" or the multigenerational family network, which is "torn" (De Vries, 2002, 307; Moor, 2004, 115).

15. An interesting example of a novel that does *not* work according to the logic of abjection is the 2009 novel *Scott Zuyderling* by Erich Zielinski, an author from the Dutch-speaking Caribbean (born on Bonaire, he now lives on Curaçao). The main protagonist's son is presented as an effeminate gay man from whom the father is alienated. The son lives in the Netherlands and appears to be married to a talented, masculine man, who applies for a job at the protagonist's office in the Caribbean. The two men will return to the Caribbean to settle there.

16. In a short story that is closely affiliated to *Aelred's Sin*, Lawrence Scott (2013) describes how two young boys take male and female movie stars from Hollywood films as guides for their own same-sex kissing and love making ("Tales," 193). A comparable logic seems at play.

17. Alluding to the acronyms FTM (Female-to-Male) and MTF (Male-to-Female) sometimes used to describe transgender identities, I propose a reading of Señorita as, in a certain sense, a Female-to-Female transvestite.

18. On the other hand, the poststructuralist concept of hybridity would be synonymous to Derridean différance, which would make it so broad that it would lose its productivity as an analytical tool. Indeed, the figure of the transvestite can—potentially—signify *any* social crisis, and advocate *any* form of reconciliation, healing, or subversion. From a different point of view, others have pointed at the successful commodification of "hybrid" life styles. Benetton is perhaps the most-discussed example.

19. Represented by new materialist (often Australian) feminist scholars such as Moira Gatens, Elizabeth Grosz, Claire Colebrook, and Rosi Braidotti.

20. In *Huis*, too, we find the recurrent insistence on the need to know. However, knowing is presented in both the Freudian sense (the desire for knowledge is inevitably also the desire for sexual knowledge too), and in a Buddhist sense (knowledge is the knowledge of life's transiency), here presented in the image of Siddharta's glance over the fence (78). It seems to me that both forms of knowledge are closely related: procreation and transience, sexuality and death are inevitably linked.

21. The ethics in the Deleuzian narrative is based on the growth of one's power and potential (Grosz, 1993, 172).

22. I would not argue for a Hindu or Buddhist reading of the narrative's ethics, in spite of, for example, the reference to the Ramayana (76) or Siddharta (78), or the clay dolls the girls play with when young (the image of the clay body being a sign of transiency in Hindu death lyrics, as literary critic Els Moor explains (2004, 114). The novel's ethics are much more syncretic.

23. This discourse seems to be closely affiliated with the Deleuzian-Glissantian discourse of becoming. Indeed, Peter Hallward (2001) has analyzed the affinity between Buddhist-inspired theories of postmodernism and postcolonialism, and other postcolonial theories of singularity, such as Glissant's. The imagery in this novel is, however, markedly different from Glissantian discourse. For example, the spiritual longing for union with nature is very strong.

24. Syncretism respects the incommensurability of the components of an assemblage, which are not seen as opposed. In this sense it differs from the notion of synthesis that is at the heart of the concept of hybridity, and that conceptualizes difference as binarism and therefore as resolvable (Becquer and Gatti, 1991).

Chapter Eight

Diversity

Writers, Multiculturalists, and Ecologists in the Netherlands

> "I am not a lover of nature, or of nature imitated, and I steered clear of the park."[1]

This sentiment is expressed by the unhappy, nervous hero and narrator of *Salomon* (2001), a novel by the acclaimed linguistic virtuoso Hafid Bouazza, a Dutch writer of Moroccan descent, who abhors being categorized as such.[2] When the hero states his lack of enthusiasm for nature, he is absolutely right. Nature, in Bouazza's dizzying novel, is dangerous—it is uncontrollably invasive. Its sheer vitality threatens the shy hero.

> In the mirages the busy town fluttered like a far-away, immeasurable clothesline. She became unrecognizable. [. . .] The Royal Palace was a miserable heap, on which the mirages played the accordion, it was a vibrating mass of living stone, it seemed to crawl with lizards and frogs who jumped up and down and to and fro on the clinkers, incessantly, while the maddening swarm of pigeons, turned into a grey-black mass, vibrated like flies on a corpse and flamed and melted like asphalt—and stank like asphalt. (44–45)

Mirages like these torture the hero, whose mental health is seriously disturbed by the end of a love affair. The very stones of the town come alive, and living beings are transformed into a vibrating mass. The oversensitive eyes of the narrator perceive both the organic and inorganic cityscape as a threatening, metamorphosing sphere of boundless vitality. In the first sentences of the novel, the narrator warns us that his extreme sensitivity borders on madness, and that the near-madness from which his eerie perception of nature stems is triggered by the presence of women and their capricious sexuality. In elaborate metaphors, the novel compares women to flowers who mutate into plantlike monsters:

> Tirelessly, the women opened the ever-refreshed flower stands of their charms, shook as hollyhocks just as many flower-bells. Just as many tall statures they shook, while at the same time they projected their thorns. [. . .] Were they freshly picked flowers with lumps of earth on their roots, or did they overnight in buckets of water? Yet nothing of this appeared capable of resisting the tortures of the mirage: many-armed, cactus-shaped monsters they became, in the glowing creek. (45)

And in the city's womb, a caged griffon lies, and the unusual "brooding heat was there to stimulate his growth"[3]—until it rises and destroys the hero's house (56). At first sight, this perception of nature as gruesome vitality seems nothing but the feverish vision of a mentally disturbed, brokenhearted young Dutch man, who struggles with his intense loneliness amidst the sexual vitality of the Dutch metropolis. However, this perception does not merely testify to individual madness. In this chapter, I will work with this mad evocation of environmental vitality to discuss some of the complexities writers must wrestle with when they want to steer clear of dominant discourses of diversity and open up a space for alternative imaginations. To begin, I will show that the above madness can be read as a perfectly sensible approach to the environment by contrasting it to prevailing Dutch environmental discourses. I will focus on the way in which Bouazza subverts the much-evoked concept of "diversity," which lies at the heart of environmental discourses (biodiversity), as well as social discourses on the multicultural, globalizing world (managing diversity).

To prepare the grounds of my argument, I want to make some preliminary remarks about the place of sexuality and interculturality in the Dutch approaches to the environment, and in this alternative imagination. Though it is not immediately obvious, this imagination of the Amsterdam cityscape could be said to express a common (cultural) outsiders' experience of Dutch culture, be they migrants or strangers to the land in other ways. The omnipresence of sexuality and the sexual assertiveness of women in this culture are the main elements to which many outsiders react (a well-known fact outside the fictional world, too: many visitors to the Netherlands, whether writers or tourists, testify to being struck by this sexual openness as well). Remarkably, in this novel, the town's sexuality intensifies and becomes threatening when the sun (presented as a visitor from abroad) causes a heat wave (38); it is tempting to interpret this as a reference to the intensification of tensions around sexuality and gender that come with multiculturalization (as sociologist Abram de Swaan [2006] argued with reference to the Dutch situation, and philosopher Baukje Prins [2004], when she pointed out the gendered nature of the new neorealist rhetoric of an influential group of Dutch right wing, xenophobic, often masculinist, politicians).[4] Within the larger context of Bouazza's work, the novel *Salomon* can be read as an elaborate effort to find words and images to capture a recurring motif through which this inescapable sexuality is explored: the motif of the evasive otherness of a desired woman who is part of an often alien, always excluding, excessively sexual culture.[5]

My point is not that *Salomon* is about the intercultural conflicts surrounding sexuality. Far from it—the novel's scope is much wider, its significance much greater than that. But I do want to argue that the imagination of the natural/cultural cityscape in the novel evokes a whole range of cultural debates current in postcolonial Western societies, including the Netherlands, in which the issues of sexuality and cultural otherness are intertwined, and that this intertwining, as we have seen, is motivated by certain basic assumptions about nature. Some of the recurring elements in these public debates are central in *Salomon* too: apart from the anxious obsession with abundant sexuality, there is fear of the untamed vitality of botanical and animal life, and anxiety about interculturality, fragmentation, and the loss of boundaries. These discourses are governed by specific, often different perceptions of nature. We are reminded of the anxious turn-of-the-century theories of race and degeneration (Edmond, 2000).

The chapters above have suggested that environmental theorizing alone contains many competing discourses on nature all based on different disciplines (from the natural sciences to cybernetics, from economics to philosophy), and all serving different scientific, social, psychological, and political interests. In the Netherlands, which will be the focus of this chapter, the popularity and dominance of these environmental theories are also formed by other political factors rather than by cogency. As Dutch environmental philosopher Jozef Keulartz (1998) and his colleague Mechtild de Jong (2002) point out, there is no sustained intellectual debate concerning the different basic assumptions about nature which govern these theories. This contradictory range of environmental theories is therefore able to furnish participants in the debate on multiculturalization and globalization with completely opposed arguments about nature and the natural. Debaters of all political persuasions will be able to find an argument to naturalize their opinions on multiculturalism. For example, some will say that fundamentalist Muslim migrants threaten the easy pluralism of the Netherlands, just as an invasive species threatens biodiversity (this view amounts to what some call ecoracism, [Bolaffi, 2003, 82]). Conversely, others may hold that diversity can only benefit from the introduction of new species or cultures.

Bouazza's work, which refers to a wealth of existing connotations about nature, can be said to privilege one recognizable view of nature over others. His work shows that certain scientifically innovative, and/or non-Western perceptions of the environment (as unstable, chaotic, unpredictable) are imaginable, and might therefore find their way to the social imagination of the general public. More explicitly than in the preceding chapters, I will now focus on environmentalist discourses, and—naively, perhaps—discuss their viability. More than a decade ago, literary scholar Dana Phillips (1999) published an essay in which he criticized the naive assumption in the emerging field of ecocriticism that it would be possible to capture the "truth of ecology" (577). I do not believe that Hafid Bouazza's *Salomon* is in the least concerned with the truth of ecology. Yet, for me, his work makes sense because it breaks with a tenacious, very dominant ecological discourse. His provocative narrative of anxious multicultural sexuality leads, independently of the work's explicit project, to a sense of nature that subverts certain hackneyed notions that were already falsified decades ago. After considering the environmental discourses addressed in Bouazza's writing, I will discuss the reasons why some environmental discourses have been embraced by society while others were rejected, and ponder the interdisciplinary relations between literature and those discourses.

However, as Phillips points out, literature is not representing ecology. Indeed, innovative writers such as Bouazza are not very much interested in representing a specific view of ecology. Instead, Bouazza's work questions and exposes the unresolved inconsistencies and tensions in and between the discourses on nature that are evoked in the debates about globalization and multiculturalization. Thus, he presents less a "truth" of nature than a productive view of culture, as a sphere shaped by sexuality and power. But in doing so, he subverts the more traditional notions of nature that inform dominant views of multiculturalism. And with that, he makes it perfectly clear that a new understanding of nature is urgently needed.

DIVERSITY 1: BOUAZZA'S BOUNDLESS VITALITY

The first of my two literary examples is Hafid Bouazza's *Salomon.* What discourses of nature is Bouazza evoking through his critical imagination of the feminization and sexualization of Dutch multicultural society? What notions of diversity does he imply?

I have already argued that Bouazza's work refers to elements in older racial theories, such as the fear of boundless vitality, but I can now suggest a more precise interpretation. Bouazza's emphasis on the invasive abundance of city life can be seen as echoing the concern with invasions and floods in public discourses on Dutch multicultural society. Nautical metaphors abound. Migrants are said to flood the country, and as Marlou Schrover already pointed out in 2002, many comments speak of the great need to control these inundations to protect the nation from these floods.[6] The anxiety that lies behind this discourse is not just related to the centuries-old Dutch fear of the flood. The assumption here is that Dutch society is a limited, closed system that will collapse under the strain of invasions of migrants. This assumption is repeated again and again in most public debates on migration. Listen, for example, to the state secretary of justice, who, a few years ago, defended Dutch and European immigration policy with the following words: "The restrictive immigration policy in Western Europe is directly related to the wish to defend the existing prosperity and to maintain political stability" (Suurmond, 1998, 99, my translation). The argument is that prosperity and stability can only be maintained within a closed system. Within such a context, metaphors like Garret Hardin's "(overcrowded) lifeboat," coined in 1974, or even that of "Spaceship Earth," first used by R. Buckminster Fuller in 1963 and Kenneth E. Boulding in 1965, seem perfectly obvious.

Many Europeans are used to thinking along these lines. This discourse goes back a long way, as Dutch philosopher Hans Achterhuis (1993) argues in an influential critique, at least as far back as the philosophy of Thomas Hobbes, and John Locke and Adam Smith after him:[7] scarcity is nature's main characteristic, its carrying capacity is not sufficient for all. In the first half of the twentieth century, this argument was elaborated in new discourses about the environment which opposed earlier mechanistic approaches. In response to the more general scientific revolutions at the beginning of the century (notably Einstein's theory of relativity, and later quantum physics, and big bang theory), which effected a break with Newtonian mechanicist science, organicist models were proposed which assumed that living organisms, rather than being machines, were hierarchically organized wholes which could not be reduced to the sum of their parts. In Europe and the United States, this approach was elaborated in theories which understood nature as a self-regulating, efficient system: climax theory and its successor, systems ecology (theories associated with, respectively, Frederic Clements and Eugene Odum). This became the predominant view in the 1970s (see Keulartz [1998], Marshall [2002], and Schouten [2001]). This holistic, antimechanicist, organicist approach did not break with the dualism of mechanicism, however. For, in its most radical form, man is seen as exterior to nature—a nature which is best left alone to develop its greatest diversity. Moreover, in spite of these radical changes in the philosophy of the environment, scarcity still appears as the crucial concept whereby the earth remains a closed system, its resources limited, its balanced biodiversity easily destroyed.

In the Netherlands, systems ecology—which was strongly influenced by the philosopher Henri Bergson's antimechanicist emphasis on life's creative vital force—developed through intense debates between writers (notably the famous Dutch writer Frederik van Eeden, 1860–1932), (language) philosophers, lawyers, theologians, and mathematicians. These advocates of the Dutch systems ecology put their trust in the tendency of every natural community to develop to its fullest potential. By this reasoning, "Man" should interfere as infrequently as possible, while nature management should devote itself to limiting the influence of civilization as much as possible. This approach is nowadays still the foundation of the Dutch Junior Nature Association NJN, and, since the 1970s especially, of the dominant form of Dutch governmental nature management.[8] A related Dutch theory of the ecosystem, which is based on cybernetics, assumes that the development of an ecosystem can be predicted, controlled, and stimulated, if the mechanisms by which it regulates itself are known. Though this approach has never explicitly guided Dutch nature management, its assumption that well-informed management will produce rich, diverse ecosystems, lies at the base of the ministerial order to produce a specific Plan for Nature Policy for every province.[9] In general, then, systems ecology became as popular in the Netherlands as elsewhere in the world, where its concept of nature and diversity can be recognized in the report on sustainability by the Brundtland committee of 1987, and in the Rio de Janeiro Biodiversity Treaty of 1992.

What is specific for the Netherlands, though, is the passionate debate that has ensued between the advocates and critics of an environmental policy that aims at re-creating a primeval form of nature that is called "new nature." The writers and activists who criticize "new nature" do so on the grounds that is leads to the artificial result of the (neocapitalist) drive to produce a commercially interesting nature.[10] In this policy, writer Willem van Toorn (2004) argues, nature is disengaged from landscape, to the detriment of both (42). The landscape loses its historical readability (and with that, its palimpsest of historic and cultural meanings—its diversity, one could say), and can no longer be understood as the result of an age-old intimate interaction between people and nature. In Greg Garrard's (2004) ecocritical terms, van Toorn and his like-minded colleagues protest the fact that an aesthetic of dwelling (as a responsible interaction with the landscape one inhabits; see chapter 6) is replaced by a commercialized version of the aesthetics of tourism (such as the pastoral and the sublime).

The urgency of this conflict is increased by two facts. First, the Netherlands is a small, heavily industrialized country that created itself through inventive and exhausting feats of engineering. This effort affected the landscape to such an extent that there is a general consensus among the public that there has hardly been any "real" nature left in the Netherlands, a consensus that exists since the middle of the last century. Second, all attempts to discuss the fields of nature protection and environmental policies are immediately read as political interventions—and therefore as self-interested party politics (van Toorn, 2004, 41). In the Netherlands, moreover, religious and spiritual readings of nature have always been influential, and there is no equally important tradition of social and historical approaches to nature to match (see below; Hoving, 2010). As social ecologists find it hard to make themselves heard, the debate is hampered. The problematic assumption that nature is a closed system, marked by scarcity, its biodiversity easily threatened by outside interference, is thriving happily.[11]

Bouazza's representation of nature emphasizes its threatening abundance, and as such it seems to echo the Dutch fear of floods. However, a careful reading shows that Bouazza's hero's approach to nature is completely different from most other Dutch discourses, and certainly from that of systems ecology. His perception of nature is anything but a balanced, diverse ecosystem, of which the scarce resources are threatened by invading migrants. On the contrary. In Bouazza's hero's imagination, nature can never be dominated. It will inevitably burst through all efforts at controlling it. In this literary universe, space is limitless; its abundant life multiplies itself everywhere. A palace becomes a mass of living stone, teeming with animal life; transformations, mutations, and incessant growth characterize city life.

This imagining of nature's abundance offers not only a perception of nature that is wildly at odds with the dominant Dutch environmental discourses, but it also embraces the notion of the boundlessness of culture: the metropolis is a space where culturally specific views of nature and sexuality collide and mingle (such as the Arabic and the European botanical imaginations of love and passion; see Graef and Louwerse, 2001), producing ever more stories, images, nightmares, and illusions. Waving away as irrelevant the idea that migrants invade and destroy the nation's indigenous cultural diversity, Bouazza's work suggests that the most pressing issues are those of gender and sexuality. The real problems of the multicultural metropolis originate in the difficulty of managing the always intense, always conflicting sexual desires, imaginations, and relations. If Western imagination has for a long time preferred to see its cultural others in the Orient as naturally passive, feminine, and sexual rather than rational, as body rather than mind, as closer to animals than to humans, Bouazza's novel subverts the original Eurocentric argument: his confused hero, sick with the pain of frustrated love, feels overwhelmed by the intensely sexual, feminine Western metropole, where human beings are constantly turning into animals and plants, and where nature always threatens to burst through the thin veneer of civilization. *Salomon* shows that multiculturalism intensifies the metropole's sexuality, increasing both this sexuality's vitality, and multiplying the conflicts and tensions that spring from the collision of emotions about sexuality and gender.

Bouazza's novel is about the tempestuous multiplicity of desire, and as such it can also be seen as a comment on diversity (of desire, of perceptions of nature and sexuality, of the appearances in which life presents itself). His work can even be said to understand unpredictable change and dynamism as the inevitable condition of the multicultural world, and thus as the sources of diversity.[12] This view stands in stark contrast to the holistic theories of nature and culture mentioned above, which hold that diversity springs from stability, and that therefore the stability of ecosystems should be protected and stimulated by being left untouched. In one sense, then, Bouazza offers a minority standpoint. But that does not mean he is wrong. It makes very good sense to read his intervention in the debate on diversity as a welcome, radical, alternative approach to the mainstream view in environmentalism. For the debate is far from settled.

Scientific arguments alone are not capable of deciding the point. If the advocates of holistic ecology are in the majority, this is not because science has proven their theories right. More and more ecologists oppose the notion that nature will always endeavor to reach a stable, harmonious balance with maximum diversity. Research suggests that ecosystems are always reacting to outside influences, and should rather be characterized as dynamic and unstable. Contrary to the belief that diversity thrives in a stable system, an ecosystem's stability can even be seen to diminish by the growth of the number of populations (Keulartz, 1998, 183). Such empirical findings are now being integrated in new ecological theories, which are influenced by modern scientific insights, such as chaos theory and evolutionary ecology (Keulartz; see also Dana Phillips, 1999, 579). Still, the holistic and organicist champions of stability continue to decide the environmental policies in the Netherlands, and beyond.

According to some critics, the main reason for their continuing prominence is because their view offers policymakers an opportunity to act (Keulartz, 1998; DeJong, 2002). For my argument about the Dutch discourses of diversity, however, another remark is even more relevant. As one critic stated with irony, systems ecology is also popular because it offers an attractive holistic rhetoric in which nature appears a perfect, harmonious, organic whole. This way of imagining nature is evoked in cultural critiques of present-day mechanized, globalized, multicultural society, where it promises to reintroduce a welcome authenticity that has been lost. Specialist of Dutch fin de siecle literature Mary Kemperink (2001) argues that the spiritual Darwinism of the era can be understood as a response to the deep wish to reconcile the desire for the transcendence of the dualism between spirit and matter with the opposed desire for scientific truth (253; see below). Her argument reveals that the contemporary desire for a unified ecological approach that is valid in science, ethics, and aesthetics had a counterpart a century earlier. Apparently, the Dutch imagination of nature is marked by the quest for a spiritual truth that would be sustained by scientific evidence.

The debate about biodiversity, then, is informed by different disciplines, professions, discourses, and interests, and this interdisciplinarity makes it very difficult to reach consensus about what is the most accurate and most productive approach to diversity. Let me give an example of a Dutch conflict between holistic New Ecology (as Keulartz calls it) and liberal multiculturalism, to show how inevitable the collision between these interests and insights is and how hard it is to decide the debate.

For some time now, influential Dutch organizations of nature conservation have been engaged in the creation of areas of so-called primeval nature. These modern wildernesses should offer the citizens of this densely populated and highly industrialized nation, an opportunity to reexperience "real" nature. An article on a related project to create a truly native forest with only native trees, was unexpectedly met with indignation. Should we not see such a project, a reader asked in an outraged letter to the editor, as inimical to society's duty to strive for multiculturality?

> The article "Alien stock" was "only" about trees, but the theory on which it is based is suspiciously close to the theory that led, in a now friendly and respected neighboring nation, to atrocities committed against human beings only some decades ago. Such xenophobia may nowadays only be uttered with restraint when it concerns people. But when it concerns nature it is apparently professed and uttered freely. "Love for all that grows and blooms" is apparently affected by the hatred for migrants and exotics.[13]

This reaction shows how much the debate on "diversity" is riddled with ambivalence.[14] While the nature conservation project hopes to counter the loss of diversity by preserving old genetic material, the reader understands diversity as the result of inviting new species and new genetic material into the ecosystem. It is clear that the conflict stems from the difference I signaled above, between understanding diversity as springing from stability, and understanding diversity as the result of dynamic conditions. But it also refers to the vexed issue about whether diversity can be managed at all. The angry reader is right in questioning the assumption that a reconstructed "authentic" wilderness, left to itself, would be the best condition for diversity. But opening up the environment to migrant species is not a guarantee for diversity either, as studies of damaging invasive species show. I do not have the space to spell out all the lessons to be learned from this modest case,[15] but I want to offer two provisional conclusions. First, the dominant view in Dutch environmental policy may hold that diversity thrives in a closed ecological system, and, for a variety of nonscientific reasons, may continue to do so in the face of evidence to the contrary. But that does not mean that the Dutch public agrees. The ways in which the Dutch public experiences nature may be different from what the policymakers assume; their experiences of nature may be more varied, and perhaps even less nostalgic. Second, this example suggests that it is very hard to talk about the management of environmental diversity without taking recourse to sociological and cultural arguments, and vice versa. This is not due to some deplorable inability to respect the boundaries of strictly separate discourses. Rather, this intermingling of discourses may spring from the actuality that these discourses are already intertwined with one another.

Jonathan Culler (1994) once suggested that new theoretical insights often occur when a theorist borrows a concept that was developed within the context of one discipline to use it in another discipline. Such interdisciplinary translations are illuminating, even if they are not the willed translations by interdisciplinary scholars. They may bring to light unexpected dimensions in the holistic environmental theories I have discussed. History offers informative examples of what happens when early-twentieth-century holistic theories are translated into insights in the nature of man and society. To put it bluntly: the holistic view has been used to defend isolationist, monocultural, nostalgic, and downright totalitarian notions of society. Keulartz points out that the holistic assumption that the interests of the community (the ecosystem) are more important than those of the individual, might explain the relation between these organicist models and destructive totalitarian ideologies, such as national socialism (Biehl and Zimmerman, referred to in Marshall, 2002, 249).

My discussion offers a less dramatic example of the close relation between the environmental discourse on nature as a harmonious, stable, closed universe, where diversity thrives, and the Dutch debates on multicultural society. But just as in the nineteenth and twentieth centuries, debaters about culture today are still reaching out to natural discourses to find arguments. Because the connection between natural and cultural discourses on diversity is inescapable, I think that the debate about multiculturalization and globalization would benefit from alternative discourses to the very seductive romantic, holistic discourses that picture indigenous nature as a bountiful stable entity which is only threatened by man's intervention. This is the very point made by Rom Harré and his colleagues in their 1999 study *Greenspeak*, about the inadequacies of environmental discourse.[16]

But where is one to look for such alternative discourses on nature's diversity? It might not be such a bad idea to look at literature. Within a literary context, the eager exploration of alternative perceptions of nature, such as those that stress the boundlessness of life, is far from exceptional. The representation of nature as excessive has received different connotations within different literary contexts, from the Romanticist celebration of nature as an unbridled life-giving force, to the anxious focus in fin de siècle writing on nature's destructive degenerating powers, which manifest in nature's immoral acts of hybrid procreation. In the Netherlands, both movements participated in shaping the literary field around the turn of the century (1900). After a short emergence in 1830–1840, Romanticism made a successful comeback in the well-known literary movement of the 1880s ("de Tachtigers"), though in a specific form, that responded to the issues of the late nineteenth century. Kemperink (2001) shows how the Dutch literary field around 1900 was generally sensitive to a spiritual understanding of evolutionary theory, a "transcendent Darwinism" (250). The idea was that life would evolve to a higher spiritual consciousness. The notion of life as a boundlessly creative force had clearly vitalist dimensions: life came to mean Life-force, a term referring to the presumed autonomous force that animated the evolution. Life was thus understood both in its scientific, and in its philosophical and mystical sense (251). It is in the latter sense that life is again conceptualized as well-ordered and striving for a state of stable perfection. The notion of life as boundless transformative energy associated with evolutionary theory is downplayed. In the literary imaginations of the period, then, science, philosophy, and mysticism were brought together to create a new, holistic, and vitalist perception of life that could boast of scientific corroboration (253), but fall back on earlier assumptions on nature's stability.

But literature also offers more radical imaginations of nature as boundlessly creative and unstable that question these holistic, organicist preferences. Bouazza's vitalism, though it is not completely new to the Dutch literary field, is much more radical than its predecessors. If we can discern continuity with that older tradition (especially in its conveyance of the anxiety associated with sexuality), Bouazza's fiction shows more affinities with more recent, post-Darwinian, as yet revolutionary voices in the debate about the nature of evolution. These new interventions bring together biodiversity studies, chaos theory, and the new evolutionary theories that break with the older tendency of explaining specific physical characteristics or behavior as immediately evolutionary useful. Bruce Bagemihl (1999), the author of a pathbreaking, provocative monograph on the bewildering diversity of sexual and affective behavior among animals, offers a seductive account of these scientific developments. Apparently, he argues, it is impossible to explain the evolutionary purpose of every single variation in color, shape, or behavior, of even of "one more kind of beetle" in a rain forest that already contains hundreds of thousands of species (252). Chaos theory shows that the issue of rampant diversity can only be understood when one replaces the focus on an individual or specific form or behavior with a focus on larger systems. Chaos theory helps to see the world as a nonlinear, immensely complex, chaotic system, which is incessantly generating "unpredictable, but not random, patterns" (247).

Diversity, then, and the "plasticity" and the "versatility" of behavior, "is best regarded as a *manifestation* of the larger 'chaotic ordering' or non-linearity of the world, rather than merely a response to it" (251, emphasis in original). Bagemihl proposes to define biodiversity as "the principle that *the vitality of a biological system is a direct consequence of the diversity it contains*" (249, emphasis in original).[17] Moreover, he wants to include "*systems of social organization*" within the concept of biodiversity (251, emphasis in original). He suggests that "sexual, social, and environmental variability may be closely allied" (251). He bases this suggestion on the fact scientists have found that variations in sexual behavior may be related to changes in the social dynamics of a certain animal community. Behavioral plasticity, as he calls it, "may strengthen the ability of a species to respond 'creatively' to a highly changeable and unpredictable world" (251). Bagemihl then turns to writer and philosopher Georges Bataille, who sees "excess and exuberance [as the] primary driving forces of biological systems" (more so than scarcity or functionality, 253), and to a wealth of chaos scientists and biodiversity theorists who emphasize life's extravagance to present his thesis of Biological Exuberance: "The challenge confronting life [. . .] is not scarcity, but excess" (253).

The source of this excess would be the immense quantities of solar energy unleashed on the planet, which would lead to a decidedly wasteful natural environment (visible in its excessive, exhausting forms of sexual reproduction, in death, and feeding practices). When Bagemihl proceeds to find parallels with indigenous cosmologies, it becomes clear that his approach might lead to a certain problematic spiritual discourse, just as his evocation of the questionably holistic Gaia-theory. However, Bagemihl does criticize New Age appropriations of indigenous beliefs and pleads for a respect for the specific, and (in nonindigenous eyes) radical aspects of these beliefs, especially when it comes to the important role played by homosexuality and transgender expression, both sometimes associated with fecundity (259). Bagemihl's study is a massive contribution to the urgently needed rethinking of the relation between nature and sexuality. However, Bagemihl's only partly successful struggle to create links between the subversive, queer insights in nature's excessiveness on the one hand, and holistic discourses on the other, shows that we are as yet far removed from a convincing new social imagination of the environment as exuberantly diverse.[18] Writers such as Bataille—and Bouazza—are as important in this slow transformation, as scientists.

One of the problems in this transition is that the new environmental theories in which diversity is linked to dynamics are hardly ever taken up in everyday speech. However scientifically sound they may appear, they do not (yet) offer an inspiring rhetoric, nor do they offer guidelines for ethical behavior. Bagemihl tries to bridge the gap between science and imagination with exhilarating prose, and the importance of style in his academic argument suggests the central role of literature in the social reimagination of nature. And yet: even if the work of brilliant writer Hafid Bouazza does offer an alternative view, and seduces through the sheer power of its language, it can hardly count as an effective, coherent counterdiscourse. This may be a disappointing conclusion to those who expect the new postcolonial and migrant literatures to do what everyday speech cannot do: offer productive new imaginings of diversity, or even an ethics. But that is just not how literature works. In contrast to those theorists of globalization who hold that we need new images to grasp the multicultural, global world, and that art might provide such images (Appadurai, 1996, 31), I want to argue that migrant writing often refuses the task of providing such imaginings. Bouazza's novel, however, offers something that is at least as valuable. By playing with different elements from a range of colonial, modernist, Arab, and postmodern, post-Darwinian discourses about nature, it exposes and maximizes tensions within Dutch multicultural society. This might be a far more relevant task than the invention of a new, conscientiously realist imagination of diversity. My second literary example will serve to argue this point in a different way—that is, that such representations are perhaps not even desirable.

DIVERSITY 2: OMBRE'S ANTI-GLOBALIST NARRATIVES OF DIVERSITY

The short stories of Surinamese-Dutch writer Ellen Ombre make their point by focusing on a completely different aspect of the debate about biodiversity: they address the economic and political contexts of globalization. In addition, they seem to build on a specific Caribbean notion of the interrelatedness of the global world. Their references to economics and politics, nevertheless, have great impact on their structure and style.

In the first story in the collection *Vrouwvreemd* (1994), the heroine's mother sighs that Europeans have so much more admiration for Surinamese nature than the Surinamese themselves. "'Those people value our nature more than we do ourselves,' my mother said when she was cultivating plants in pots and complained how hard it was to obtain large Planta-margarine tins for her gerberas" (8–9). But the father shrugs. "Who cares. Nature! We were taught not to love this country" (9).[19]

And he goes on to argue that Europeans often soon tire of their initial aesthetical enthusiasm. This conversation is interrupted by the narrator's comment about the mother's activities, busying herself with her home-grown plants. For that purpose, as the quote shows, she once used the big tins of a certain brand of margarine. However, this margarine is now no longer for sale, after it apparently all but poisoned whole families. The spoiled product had been dumped in Surinam by the Netherlands, a depressingly familiar destructive commercial strategy that marks the relations between North and South. The mother now complains that the new butter tins are too small for her cultivating needs. Her effort to create natural diversity in her little potted garden is frustrated by the Surinamese resistance to exploitative global economics.

In this ironic manner, Ombre locates the debate about nature in the context of global trade. Thus, a new aspect of the debate about diversity comes to light. Biodiversity has nowadays become an economically quantitative notion instead of a qualitative concept. It is conceptualized and traded as a resource; old genetic varieties are turned into market commodities (see, e.g., Weizäcker [1993]; Franklin, Lury, and Stacey [2000]). Nature is thus changed from boundless diversity into a provider of marketable resources, and thus, scarcity is created. However, the example of the mother's need for the global market to create biodiversity shows that there is no easy way out. Nostalgic musings about a return to authentic nature would be a particularly useless response. The popular holistic discourses on biodiversity cannot serve as an alternative, because these depart from the assumption that an ecosystem is a balanced whole. Nevertheless, the desire-inspired projection of coherence onto nature's diversity is very close to the desire for (economic) unification in the discourses and the strategies of the global market that Ombre criticizes.

Even if Ombre's stories are concerned with exploring cultural diversity, and the need to find better ways to manage multiculturalism, her narrative style questions the desire to find a coherent discourse by means of which to represent the diversity of multicultural society—to the point of questioning the possibility of representing this diversity at all. This is made explicit in a story about a young Hindu girl from Surinam who marries in Holland in the middle of a confusion of desires, fears, cravings, rituals, consumption goods, and status symbols from all over the world. One sentence is repeated in this short text, narrated in fragments: "It is not easy to capture everything in one image. It is too much at once" (101).[20] Ombre's fragmented narrative echoes this as if it were a writer's statement. I read this reluctance to create a coherent image as a critique of unifying discourses of diversity, such as the dominant discourse of globalization mentioned above. For the diverse elements of the multicultural experience can only be brought together in one image when they are treated as quantitative entities. And that is of course precisely what happens in the global market that Ombre criticizes. Polemically, Ombre's aesthetics emphasize the incommensurability of all these events and things. Cultural diversity cannot be quantified. With this statement, Ombre addresses and criticizes the dominance of the economic discourse of diversity (including its key concept, scarcity). In addition, she shows that even the art of literary writing is implicated in this economical discourse, and thus cannot but respond (however critically or unsatisfactorily) to its unifying aesthetics.

Let me clarify that Ombre's work is not just responding to dominant political and economic discourses and narratives. Her response is inspired by the Caribbean (and, more specifically, Glissantian) idea of the interconnectedness of all spaces and identities in the world, which does not lead to a harmonious whole, but nevertheless can be understood only after accepting the untranslatable complexity of every specific local place or identity, which is shaped by its relationship to other spaces. Such incommensurable plurality can never be represented in its totality. Ombre's narrative style also seems to be inspired by this notion of an incommensurable plurality. However, her understanding of this plurality is not built on phenomenological notions of the irreducible, nongeneralizing, infinite character of totality (as in Levinas, and to a certain extent also in Glissant), but rather on a political critique of the reductive and generalizing logic of the global market, which translates all qualities into quantities. The economy and politics are therefore ever-present dimensions in Ombre's writing. Her writing interweaves a mother's nostalgic exclamations about nature with stark economic analysis, and with reports of (neo)colonial bluntness and migrant naiveté. Thus, Ombre's stories show that no discourse is able to capture the painful complexity of globalization, and that no discourse can escape this intertwining.

CONCLUSION

The work of Hafid Bouazza and Ellen Ombre clarifies the unique role literature can play in this complex of intertwined discourses about diversity—it can be said to be the privileged third discourse in addition to the environmental and political discourses. Literature shows new ways of exploiting old discourses on nature. By picturing the Western metropole as a sexualized, threatening, but endlessly vital space, Hafid Bouazza evokes colonialist, (post)modernist, and certain classically Arab ways of using natural images to address issues of sexuality. In this way, he implicitly criticizes the restricted view of nature as scarce, limited, and predictable, as advocated by influential Dutch environmentalists. In addition, he offers a different approach to multiculturalization by bringing out the crucial role played by sexuality and gender in the conflicts of the multicultural metropole. Ellen Ombre combines her sharp critique of economical discourses of biodiversity within an aesthetics of fragmentation that counters the ideology of coherence created by the global market, or the naive embrace of connectedness in some postcolonial studies. At the same time, her aesthetics seem to be inspired by a Caribbean poetics of diversity through which she can imagine the non-totalizing connections between all incommensurable elements of globalization. These writers do not offer a recognizable, coherent new environmental imagination. But their readers are rewarded by a clear insight into the conflictual entanglement of the different discourses about diversity, and into the many gaps in the Western-European and North-American ability to imagine the multicultural cultures and natures of globalization.

NOTES

1. My translation of: “Ik ben geen liefhebber van de natuur, noch van imitatienatuur en meed het park” (44).

2. Hafid Bouazza came to the Netherlands from Morocco as a young child, but he detests being pigeonholed as a Moroccon or migrant writer. While migration and multiculturalization are important themes in his work, it is too rich to be reduced to the genre of “migrant writing.”

3. My—again tentative—translation of: “De vrouwen openden onvermoeid de immer verfriste bloemenstallen van hun charmes, schudden als stokrozen zovele bloemklokken, zovele rijzige gestalten, maar staken tevens hun doorns uit [. . .] Waren zij versgeplukte bloemen met aardkluiten aan de wortels of overnachtten zij in emmers water? Maar niets van dit bleek bestand tegen de martelingen van de mirage: veelarmige, cactusvormige monsters werden zij in de gloeiende kreek” (45), “broedende hitte was er om zijn groei te bevorderen” (56).

4. My tentative translation of Bouazza’s incomparable prose. The original: “In de luchtspiegelingen wapperde de drukke stad als een verre, onmetelijke waslijn. Zij werd onherkenbaar. [. . .] Het Paleis was een miserabele hoop, de luchtspiegelingen speelden er accordeon op, het was een trillende massa levend steen, het leek te krioelen van hagedissen en kikvorsen, die op en neer en heen en weer op de klinkers sprongen en dat onophoudelijk, terwijl de gekmakende zwerm duiven een zwartgrijze massa werd, trilde als vliegen op een kadaver en vlamde en smolt als asfalt—en stonk als asfalt” (44–45).

5. The character comes most explicitly to life in “Apollien,” the story about the fated love affair between a young Moroccan man and a confident Dutch girl in Amsterdam.

6. In *NRC Handelsblad* 8/9 June 2002.

7. The assumption that scarcity would be nature’s main characteristic, and that stability can only be maintained in a closed system has been questioned by critical thinkers such as social ecologist Murray Bookchin and philosopher, theologian, and pedagogue Ivan Illich. See Achterhuis (1993) and Keulartz (1998).

8. DeJong (2002) and Keulartz (1998); see also Schroevers, 1999, 142–45, and van der Meulen, 2009, 245–48; with some useful nuances.

9. DeJong, 2002, 284.

10. Its success is visible in, for example, the unexpectedly great popularity of the awarded Dutch nature film, followed by a television series, *De Nieuwe Wildernis* (The New Wilderness) (2013), directed by Mark Verkerk and Ruben Smit. (http://www.denieuwewildernis.nl/)

11. For an account of other reactions of the general public, see van der Meulen, 2009, 296–99.

12. Which does not mean Bouazza’s work is based on chaos theory or any other environmental theory. His work should rather be understood within a transnational context of classical, biblical, and medieval Dutch, European, and Arabic cultural traditions.

13. *NRC Handelsblad* August 17/18, 2002, 33 (Hester van Santen. “Vreemde stammen; kweekprogramma moet autochtone bomen redden.”) and August 25, 2002 (H. A. Vonder, “Exoten 2.”) My translation of: “In ‘Vreemde stammen’ [. . .] ging het ‘slechts’ om bomen, maar de theorie waarop het berust leunt wel zeer dicht tegen die welke bij een thans bevriende en geachte buur-natie slechts enkele decennia geleden leidde tot gruwelijkheden mensen aangedaan. Die vreemdelingenhaat waar het mensen betreft mag dan tegenwoordig in beperkte mate geuit worden, waar het de natuur betreft is het kennelijk in bepaalde kringen iets dat beleden wordt en vrijelijk tot uiting komt. ‘Liefde voor al wat groeit en bloeit’ is kennelijk aangetast door allochtonen—of exotenhaat.”

14. In the second decade of the twenty-first century, the remark is almost atypical; most of the Dutch audience’s responses today would express a vehement critique of multiculturalism and antiracism.

15. Such as the observation that, in general, invasive species become pests most often if the receiving ecosystem’s diversity has already been destroyed, and turned into a vulnerable monoculture (as in large-scale agriculture), see Bright (1999). Let me add that the letter-writer is not right in accusing the author of the article or the advocates of the project of xenophobia or racism. Indeed, one of the speakers quoted in the article rather tries to undo the xenophobe dichotomy between natives/aliens, and the other speakers offer a variety of views on diversity.

16. Julian Walker makes the same point in his 2003 essay "The Language of Encroachment."

17. One might want to criticize the following, more problematic statement in which Bagemihl (1999) relates the increase of diversity to the increase of stability in an ecosystem; see below in the main text.

18. See Ladelle McWhorter, for a thought-provoking queer critique of the concept of diversity, based on its problematic genealogy.

19. My translation of: "'Die mensen hebben meer waardering voor onze natuur,' zei mijn moeder toen die planten in potten kweekte en zich beklaagde dat het zo moeilijk was om aan grote Planta-margarineblikken te komen voor haar gerbera's [. . .] 'Ach wat, natuur, het is ons afgeleerd om van dit land te houden'" (8–9).

20. My translation of: "Het is niet eenvoudig alles in beeld te vangen. Het is te veel tegelijk" (101).

Chapter Nine

Opacity

On the Dutch Myth of Openness

Time and again, people who come to the Netherlands are struck by the pronounced openness of the Dutch landscape and the Dutch people. The hero in a semiautobiographical novel by an Iranian in exile sighs: "I had to get accustomed. [. . .] Thus, I had to get accustomed to the dominant color green. A cow in the mist was new to me. Those bare legs, bellies, breasts, buttocks and the language were all things to which I had to get accustomed" (Abdolah, 1997, 13).[1] But not only people bare themselves. The renowned Dutch landscape, with its wide skies and flat meadows, is sometimes even seen as open to the point of being transparent. Nothing stays hidden in such a space. The author of the above mentioned novel remarks, with irony, like many immigrants and visitors before him, that Dutch people boast of having created this territory themselves; they live in a transparent space that they think they know and control completely (Blerck and Harsema, 2001).

Openness, a strong belief in visibility and directness—the need for, and the belief in, control and regulation—these are a few of the characteristics that may be found in many representations of the Dutch. They suggest a realistic, rational, and sober culture that sustains a general acceptance of things that would remain hidden elsewhere. A 2004 survey showed that the Dutch still believe that soberness and common sense are their primary national characteristics, and that they characterize their own individual identity as primarily tolerant. Dutch immigrants defined themselves in the same way, disagreeing only slightly with regard to their soberness. Moreover, the Dutch express the opinion that the rest of Europe should adopt these Dutch values as well. The values of tolerance, love of freedom, and soberness have been mentioned repeatedly by the architects of Dutch national identity, such as

Siegenbeek in the early nineteenth century, Potgieter and Bakhuizen van den Brink in the 1840s, and Huizinga and Romein in the 1930s and 1940s. Present-day public debates about Dutch national identity reiterate these primary values of Dutch culture: tolerance, consensus, egalitarianism (Scheffer, 2000). Different narratives on the historical origins of these values exist. A relevant narrative for the context of this study, is that these values have been analyzed as stemming from the typical conditions of the Dutch environment: as inhabitants of a region threatened by the sea, and wrested from the sea, people had to set aside their differences and coöperate for the sake of survival. The social values of tolerance, consensus, and egalitarianism are thus naturalized as inevitable responses to environmental challenges. The Dutch term "poldermodel," which refers to the model of negotiation—instead of open conflict—to settle differences, is a good example. It evokes the ecological site that is perhaps the most characteristic of all Dutch landscapes: the polder, the hard-won result of human intervention in the Dutch environment. It refers to social and political insitutions, while at the same time depoliticizing them by suggesting their origins in a prepolitical past, when a community would be formed in the shared struggle against nature. On the other hand, the term "polder," which explicitly refers to the social constructedness of this environment, also seems to invite, against the grain, a social and political reading.

In this chapter, I will explore the interaction between the ecological and social/political dimensions of the related set of notions of openness, transparence, and tolerance. My leading question will be: to what extent do these environmental/social notions contribute to the understanding and management of multiculturalization and migration?

At first sight, one might answer: to a great extent. The Dutch tradition of tolerance has long been an inspiration to other European countries in search of an effective immigration policy, perhaps most especially to Germany, with which the Netherlands shares kindred problems in their transition to a multicultural society. However, in contrast with Germany, the Netherlands has had an integration policy in place since the beginning of the 1980s. In 1994, the Dutch government praised its own immigration policy as exemplary; it was regarded as successful and in advance of other European nations (Böcker and Tränhardt, 2003, 3). One of the main causes of the success of the immigration policy was believed to be its formally delineated tolerance of cultural difference—while this encoded respect for cultural identity might be deemed the main element lacking in the German management of immigration.

Only a decade later, however, the Dutch success was seriously questioned, the debate about immigration and multiculturalization soured, and there was ample reason to (re)consider the nature and effect of this renowned Dutch tolerance for cultural difference. In the following, I will unravel the rather baffling knot of national values and characteristics, in which tolerance can carry opposite meanings, through a consideration of the work of environmental activist and immigrant authors. The question is all the more relevant as, in Dutch public debate, the common assumption is that an emphasis on openness and tolerance would facilitate integration. The fact alone that most theories of migration emphasize the complexity and opacity in most immigrants' artistic efforts at connecting to their new cultural surroundings should lead us to question that allegation. If openness and tolerance are sufficient conditions for immigrants to inscribe themselves in a new cultural space, why do so many scholars point at the complexity, layeredness, opacity, heterochronicity, or mosaic structures in their writing (Bal and Hernández-Navarro, 2012)?

This chapter's narrative moves from a debate of openness to the appreciation of the emergence of opacity as a much needed alternative concept. In the course of the chapter, it will become clear that a critical analysis of this debate cannot satisfy itself with a critique of society; it must include a critique of Dutch environmentalism. First, however, I will sketch a modestly comparative social and political context.

IMMIGRATION POLICIES IN THE NETHERLANDS

Up until 2001, most of the autochthonous Dutch certainly believed that the Dutch integration policy was a success. The break came with two sets of events that occurred in 2001, 2002, and 2004: first, 9/11; second, two political murders. The first murder was the shocking slaying of Pim Fortuyn, a flamboyant, gay, right-wing politician, who had been questioning the multicultural society in unusually sharp, populist terms. The second was the brutal killing of filmmaker Theo van Gogh by a young fundamentalist Moroccan-Dutch Muslim in 2004. Van Gogh was the director of Somalian-Dutch politician Ayaan Hirsi Ali's film *Submission*, an incisive critique of Muslim attitudes toward women. The reaction to the first murder in 2002 (by a white animal activist) was without precedent. Overnight, the whole nation seemed to change its opinion about the successes of the Dutch policy of immigration and multiculturalization. The new consensus among white people seemed to be that the much-praised policy of the 1980s, that is "integration while retaining one's cultural identity," had encouraged immigrants to stay within their own communities, unemployed, engaging in criminal and antisocial activities, and clinging to their cultural values, which now appeared deadly. The second murder reinforced these sentiments.

In hindsight, the change is not really surprising. The positive evaluations of the Dutch policy were often based on a consideration of only the formal legal status of immigrants, neglecting other aspects. As in France and England, it has been relatively easy to obtain national status and equality before the law in the Netherlands—especially when compared to her close neighbor Germany. In addition, the Dutch antidiscrimination laws are strong, and minorities have been granted more cultural group rights than anywhere else in Europe. These three characteristics all compared favorably to German policies, especially those in effect before 2000 (Koopmans, 2003, 46). A critical assessment of other indicators of the success of immigrant integration, however, should also take immigrants' position in the labor market into account. Detailed comparative studies and discussions between researchers lead one to conclude that, in this respect, the position of immigrants has long been worse in the Netherlands than in Germany (Böcker and Tränhardt, 2003; Koopmans, 2003).[2]

The Dutch tolerance for cultural difference, which was the hallmark of its integration policy until the end of the 1980s, appeared to fail as a motor for integration. So did the policy that followed in the 1990s, which replaced the respect for cultural identity with the pursuit of diversity, but which entailed a comparable reification of cultural identity. This doesn't mean that the multicultural society has irrevocably failed. The official governmental reports on multiculturalization argue that many immigrants have done well for themselves. Remarkably, they also argue that this was in spite of, rather than with the help of, Dutch immigration policy, which in their eyes has been incoherent and rather unsuccessful, especially in the social and cultural domain. A study by Tom Duyvené de Wit and Ruud Koopmans suggests that, a decade ago, Dutch immigrants tended to identify more strongly with their new nation than the German immigrants (Duyvené de Wit and Koopmans, 2001). At the time, these nuances were lost on those who expressed their criticism of the dominant Dutch immigration policy in the harshest terms—and who set the tone for the debate in the years to follow.

TOLERANCE REVISITED

Let us now take a closer look at this Dutch tolerance, which is at the heart of the Dutch integration policy. Where does it come from? What has it done, and what does it do now, within the multicultural society? Can it function as a strong motor for integration? I will here present the accepted political answers to that question.

From its beginnings as a republic onward, the Dutch nation has chosen a strategy of nonintervention to regulate the differences between its large minorities. This strategy is built on the Dutch plea for freedom of conscience and religion, which, in the sixteenth century, was directed toward the Spanish occupier, Philip II. Because the religious minorities grew large enough to prevent their marginalization during the centuries that followed, and none developed into the dominant minority, the government of the Dutch consociation (as political scientist Lijphart calls it) allowed its minorities to create its own institutions. At the turn of the nineteenth century, a further step was taken: all citizens were granted formal civil rights. This did not mean that their religious identities were acknowledged, but that the differences between their religious and political ideologies were regulated in a pillar-like structure, where each pillar was granted full sovereignty.

Dutch sociologist van Ginkel concludes that (from 1780–1790 onward) Dutch national identity did not come into being "through state intervention, by the repression of national differences," but by the "networking and 'consensual practices' [. . .] of the political-cultural elites who emphasized and institutionalized differences" (van Ginkel, 1999, 33). That is why one could say tolerance is not a consistent Dutch value, but that it exists, first and foremost, as an institutionalized structure.

Since the sixteenth century, many debates and studies have addressed this renowned Dutch tolerance (Gijswijt-Hofstra, 1989). The first great debates were a reaction to the violent religious intolerance of the Spanish occupiers in the sixteenth and seventeenth century; in the twentieth century the debates were reacting to the Holocaust. In the latter debates, the ambivalences inherent to the concept of tolerance received more and more critical attention (Rogier, 1965; Bovenkerk, 1976; Kossmann, 1984; van Doorn, 1985; Dubbelman and Tanja, 1987; see Gijswijt-Hofstra, 1989, 16–24). At the beginning of the twenty-first century, in the harsher debates in the wake of 2002, one is struck by a curious consensus between the right-wing and left-wing critiques of the concept. Right-wing speakers, the so-called "new-realists" (Prins, 2004), interpret Dutch tolerance as a national silence around the severe problems of immigration, and as the cowardly, politically correct evasion of difference, as opposed to an openness toward difference (see, e.g., Scheffer [2000], Schoo [2000]). More progressive commentators offer a comparable analysis. The Dutch sociologist and historian, Hondius, for example, argues that the passive form of tolerance dominant in the Netherlands and known as "gedogen," which means "to put up with, to tolerate a practice one disagrees with, or which is against the law," is close to the passive form of intolerance, evasion, which is one of the most prevalent forms of discrimination (Hondius, 1999, 9). Another sociologist, Cas Wouters (1990), emphasizes that the strategy of evasion in general is a Dutch characteristic. Both strategies, tolerance and evasion, are considered to be based on the radical differentiation between self and other, and they both testify to a strong sense of superiority, from which stems the authority to tolerate or evade others.[3] The two groups, however, differ in their opinions about the remedy. The new-realists propose active intolerance as a remedy: the reaffirmation of Dutch national identity, assimilation, and exclusion of those who do not conform (Scheffer, 2000; Vuijsje, 1986; see Prins, 2004, 37). One is reminded of a speech by the European statesman, Bolkestein, at Humboldt

University in Berlin in October 2004, praising the celebration of a German Leitkultur as a remedy. The statements of the new-realists are made in the name of that other strong Dutch tradition—openness, freedom of speech, and clarity. Their opponents, however, often opt for a strategy of active tolerance— the path of negotiation and interaction. Hondius offers the following helpful overview:

> Active tolerance—acceptance after active argumentation
> Passive tolerance—putting up with, overlooking
> Passive intolerance—evasion, ignoring, keeping off
> Active intolerance—exclusion, elimination
> (Hondius, 1999, 9)

Traditionally, the Dutch seem to have only two main strategies at their disposal for dealing with a variety of social, religious, economic, political, and other differences: the passive forms of tolerance and intolerance.[4] A short example may help to illustrate these Dutch ways of dealing with difference. It concerns a very curious debate that took place in April 2004. A government-supported environmental foundation found that immigrants seldom visit the Dutch forests for leisure. After consulting a few immigrants, the foundation decided to create a forest in which, contrary to Dutch habit, people were allowed to stray from the paths, pick fruits and nuts, and so forth. In this way the project would avoid the logic of exclusion that structures Dutch forests: the exclusion of culture (and people) from the domain of the natural, and the exclusion of immigrants. One initial reaction was a neutral report in a populist newspaper (*De Telegraaf*), which coined the word "smulbos" (yum-yum forest). In the wake of that article, an unexpectedly large number of people seemed to respond to the project in a completely different way than intended. It was immediately read as an ethnic (or, by that Dutch synonym, cultural) project. In an extended exercise in new-realist openness, people protested that tax money was once again being wasted on a ridiculous project that favored immigrants over the autochthonous population. White people routinely assumed that the immigrant's culture (here imagined as a homogeneous entity) dictated that "the" immigrants must enjoy nature in a culturally specific way that would be irreconcilable with that of the natives. The binary oppositions "them/us," "body/mind," "smell and taste/look" and "primitive/modern" were revived.[5] The effort to create cultural diversity within the Dutch landscape was met with aggressive protest, and with an actively intolerant plea for exclusionary homogeneity.

Another project put into effect by the same hapless inventor of the forest project might seem to be its logical counterpart. The essential aspect of this project, the Empty Spot, can be described by saying that its director dispossessed a piece of the Netherlands. By entwining an impenetrable thorny hedge, he closed off an area from the Dutch landscape, which is otherwise governed by an endless series of rules and prescriptions. There was no other way to formally liberate a piece of Dutch landscape: "For a tree it is impossible not to be owned by someone" (*De Volkskrant*, July 28, 2004; Slobbe, 2004).

The second case, despite its poetic resonance, does not represent a form of tolerance entirely different from the first case. The effort to remove a piece of nature from the Dutch landscape is, in a way, a testament to the culture's inability to engage with difference. Instead of thinking through the complexities of what a nondestructive interaction between human beings and a partly natural, partly cultural landscape would look like, the imaginative inventor opts for radical differentiation. In this, he adheres to the first-mentioned Dutch strategy of representing difference as absolute. "Inclusion, purification, branching off, exclusion—these are tenacious tendencies within Dutch culture," sociologist Gowricharn comments somewhat sarcastically (1998, 103).

It would seem that Dutch history doesn't offer many solutions to the very real need to acknowledge differences. The tenacious habit, of which one already finds examples in the nineteenth century, of understanding differences between citizens (in personal conduct, in art) as defined by culture and therefore as absolute and unchangeable, is not very effective. It leads to evasion and the institutionalization of difference.

The Problem with Openness

Several renowned Dutch immigrant writers have questioned this reification of cultural difference. At the same time, they question the assumption that Dutch tolerance has created an open, egalitarian society, and that this openness would facilitate the integration of newcomers. This critique can also be found in their imagination of the Dutch landscape. "She looks outside. The sky is leaden grey, as the water is. Her eyes are unable to pierce it" (Roemer, 1997, 65). In this short quote from a much-praised novel by Surinamese-Dutch writer Astrid Roemer, a Surinamese woman sits and watches the Dutch winter landscape, in all its forbidding starkness. Watching it, she finds that she has great trouble in *seeing* it. When she finally leaves the village, she must admit that she has not been able to see it at all. This seems to be at odds with the conviction that the Dutch landscape is openness, and visibility itself. But like Roemer, Moroccon-Dutch writer Abdelkader Benali problematizes that view:

> For me, the Dutch landscape is only sky. [. . .] In my village the world began behind the mountains. You couldn't see it, you had to imagine it. Here in the Netherlands the landscape is flat, you can always see the horizon, there is

> nothing to imagine. [. . .] A Dutch writer describes the things he directly sees. [. . .] My eyes are focussed on the things you don't see. (Bodegraven, 20)

The excessive visibility of Dutch landscape has a curious effect: it hinders the search for meaning. For writers like Benali, only indirectness is meaningful. Iranian-Dutch Kader Abdolah (1997) chooses remarkably similar words to express his views on the Dutch landscape: "Everything is open, for all to see, on the surface. The directness of flat land. You ask a question and you get an answer. That is something we don't know. In our country the excitement, complexity and mystery remain. Magic!" (8) With some nostalgia, he states that Persian poetry rather praises the beloved as "a secret behind a rock." The Dutch landscape is either excessively open and direct, or excessive in other ways, such as in its wetness: where a Persian poet will speak about the one raindrop that slowly traces its path over a girl's neck and bosom, the Dutch rain will immediately soak her, in a much less subtle way. "In this country, no birds dare fly. It rains, rains, rains" (Blerck and Harsema, 2001, 8). Both absolute openness and excess make it hard to find words for intimate experiences (Blerck and Harsema, 8).

Roemer's statements on landscape echo those of Benali and Abdolah, though they refer to a different, tropical nature: "In the tropics much is left to discover, it is mysterious, the landscape is terribly present, is always changing" (Bodegraven, 21).

That these remarks do not just testify to the lighthearted taste for mystery becomes clear when Moluccan writer Frans Lopulalan adds: "If I feel affinity with the Dutch landscape . . . then there is always the additional feeling: but it isn't really mine. It is theirs" (Bodegraven, 22). Roemer emphasizes a comparable distance to the landscape: "I describe the [Dutch, IH] landscape, but always from the perspective of a somewhat uprooted character. A character that experiences alienation in the landscape, or amazement, but sometimes comfort too" (Bodegraven, 22). The empty openness of Dutch landscape, then, comes about by its exclusiveness: it doesn't offer newcomers any refuge, no welcome sheltered spaces to linger. Its lack of complexity and curbs can be seen as hostile to plurality and variety.

Nevertheless, migrant writers often relate less to people than to the landscape as the privileged discourse through which they wish to come to terms with the Netherlands as a possible new homeland. This need to relate to the Dutch natural landscape is understandable; one cannot tell a story without a sense of place. Space is not just the setting for plot, space can be said to *be* plot, as Bakhtin's (1988) work on the chronotope showed long before ecocritics began to make that point. In this vein, Dutch landscape can be seen to offer a very particular sense of stories and plot, and it may be very hard for migrants to relate their life stories to the Dutch narrative of transparancy and emptiness.

Migrant writers are not the only ones to deplore the emptiness of Dutch landscape. If this emptiness has once been, and still is, a quality that is much praised by the Dutch itself as transparent openness (an openness and freedom which would reflect supposed Dutch tolerance and open-mindedness), it is also translated as an emptiness with very different connotations. Some time ago, Dutch philosopher Ton Lemaire argued that the rise of the natural sciences and rationalism has produced an industrialized, mindless landscape, from which meaning and mystery have disappeared (1970). Lemaire's analysis of the Dutch landscape as an empty, disenchanted space has been widely embraced, and his views have even been adopted for Dutch projects of nature management. But this pessimistic view has also been met with serious criticism.

Philosopher Hans Achterhuis (1993) states that Lemaire is wrong in assuming that the Dutch landscape is empty. Following Simon Schama (1995), he points out that the myths that animate landscape do not have to be reinvented from scratch, as Lemaire states. The cultured landscape is riddled with all kinds of meanings and myths, as the many people who enjoy hiking and cycling through it show. They are just no longer the Romantic myths about origin and purity, but rather the diverse narratives of a (post)modern society.

The dreams of origin and purity are, however, decisive to present-day Dutch environmental policies. In chapter 8, I argued that the pessimist view that there is no longer any "real nature" in the Netherlands was reinforced by a tenacious environmentalist tradition in the Netherlands, systems ecology. Shaped since the early twentieth century in interdisciplinary debates, the Dutch version of systems ecology is still dominant today, in spite of the fact that it has been convincingly falsified as a scientific theory. Its pastoralism, however, is seductive to a general audience, and its message that *nature is best left alone* offers policymakers a welcome strategy for environmental protection (De Jong, 2002; Keulartz, 1998). As a consequence, Dutch nature developers are now pursuing a Romantic landscape ideal, and focus on constructing a primeval landscape in well-defined areas within the technologized cultured landscape. These efforts to revive a fully autochthonous landscape do not seem of much appeal to migrants.

Several migrant writers aim at answering questions that are slightly different, but as urgent as Lemaire's and Achterhuis's: How can the flat, shallow and one-sided openness of Dutch landscape be experienced as a space where the complex, varied multicultural population can intimately relate to? How can the Dutch landscape become a space that welcomes its migrants' life stories? Their answers are as stimulating as they are sobering. The above quotes suggest the response that the much desired significant complexity, that allows for an identification with the landscape, is already present, and *readable* (van Toorn, 2008), in the cultured Dutch landscape, in spite of the opinion that its historical meaning is irrevocably erased.[6] In the next section, I will show how this complexity is (re)created.

Ton Lemaire is right, however, when he suggests that the narratives on the Dutch landscape's historical, intercultural complexity are not necessarily part of Dutch cultural memory. In the dominant Dutch discourse, nature is still placed outside history. It is seen as either absent from Dutch culture, or as absolutely other to Dutch culture: as spiritual transcendence, or as prehistoric wildness. It is situated exactly in the space where the Dutch desire for transparence, openness, and freedom from the complexities of history resides.

AGAINST TOLERANCE, FOR OPACITY

Dutch immigrant writers do not only question the anxious appreciation of Dutch openness as a national virtue, but they also question two related political assumptions: (1) that Dutch tolerance has created an open, egalitarian society, and (2) that integration is therefore predicated on minorities' knowledge and adoption of Dutch cultural history and values. In their responses, they open up a new discursive space in which issues of difference, openness, tolerance, and knowledge play completely different roles. Their work produces a new sense of space, which is characterized by some of the qualities that are often said to mark immigrants' art: notably, a certain thickness, stickiness, layeredness, or opacity, produced by an inventive manipulation of time, memory, and fictional intertextuality (Bal and Hernández-Navarro, 2012). Thus, they create new possibilities for belonging within a nation at odds with its own plurality.

Kader Abdolah

As an immigrant, Iranian writer Kader Abdolah has begun to love the Dutch landscape, in spite of his criticism, but he doesn't buy its nationalist myths. Confronted with the cherished remark that the Dutch have their pride of place because of their feeling that they have created their country themselves, he bluntly states that that is just nonsense. "You just have a large ego. We have also created our own world with dykes, canals, cities and cellphones. Typically Dutch to think that you are the boss of nature" (Blerck and Harsema, 2002, 8). Instead of sharing this nationalist need for control, in a story in his first collection of short stories, he adopts a curious strategy of *doubling* the Dutch landscape, in which he mingles the Iranian landscape of his youth with the Dutch landscape of his exile. He writes: "Sefiedganie, the river of my childhood, has changed her course. On this moment, she flows through the river IJssel." In the waves of the Dutch river IJssel, he sees the snakes that swam in the Iranian river; between the cows on the river's shores, he sees the child rapist who terrorized the village of his youth (1995, 16, 17). In short, simple words and sentences, that are reminiscent of the Dutch nature poets he admires, he succeeds in evoking a doubled Dutch landscape, which, in spite of its transparancy, has received a new, Iranian dimension which allows him to relate to it and to consider it a significant context for his life.

The quotation with which I started the chapter is taken from *De reis van de lege flessen* (1997), Abdolah's first novel, which continues the strategy of doubling, in spite of its fascination with Dutch openness. In meticulous, sparse, almost transparent prose, the novel narrates the story of an Iranian refugee, Bolfazl, who tries to relate to the open and permissive Dutch society in which he has landed. His closest acquaintance is his neighbor René, a gay, unemployed painter, whom he one day espies lying naked in his inadequately fenced back garden. In Bolfazl's effort to find a place in a society, which, in spite of its openness, doesn't offer many opportunities for socializing, he decides to let René be his guide and mediator. "René was like a bead in a string of beads that has been broken" (51).[7]

When Bolfazl understands that René is suffering from severe psychological problems, and that he has always been a stranger to society, as Bolfazl is now, he begins to see René as his double. "Me? I went there to see everything. To track the process. To find out how it could go when one is exiled" (59).[8] A curious game of voyeurism, imitation, and doubling ensues. "When I saw that René could no longer adequately remember his past, that he no longer knew that I was Bolfazl, I suddenly realized that I could no longer summon up that scent [of his past, IH] either" (74).[9] No more than Bolfazl's, René's life can be understood as a coherent narrative with a stable plot, energized by a strong desire. René's desire for men, far from being an acknowledgment or fulfilment of his identity, is ultimately empty, as it is frustrated by his much more fundamental, painful alienation from life. However, the refugee can take René as his guide precisely because he does not seek identity or agency. It would be a mistake to suppose that Bolfazl models his identity on René's. What he seeks is understanding—not of Dutch society, but of exile, an experience that, he realizes, is also at the heart of Dutch society. "I had to look properly, listen well, and remember everything. One day I would be able to explain what it means to be exiled" (91).[10]

Instead of surrendering to the logic of the Dutch multicultural nation, with its demand for a recognizable, fixed cultural identity, Bolfazl seeks to adopt the logic of the flight, as he calls it, or, in other words, the logic of a life in exile, and he recognizes this same logic in the life of some of the Dutch. Bolfazl creates meaning by creating a multiple referential network, in which Dutch events are understood by comparing them with Persian stories, and vice versa, while knowing full well that all those stories and insights are partial and provisional, partly true, and partly fictional. "I am looking for a correspondence between events. [. . .] To be honest, I do not really trust my own memories anymore. I invent those stories myself" (55).[11] "For an exile, fantasy and reality are intermingled" (108).[12] This semiotics is based on the index rather than the symbol. From this radically comparative perspective, transparency, directness, control, and realism become less relevant qualities in Bolfazl's aesthetics. "Stories have their own laws. One is not able to change the course of stories" (106).[13] Meaning is produced in the overlapping and mingling of stories.

Abdolah is creating an opaque sense of place, by a feverish calling forth of the many memories and fictions that seem somehow connected to the events in his new surroundings. In Mieke Bal's account of migratory aesthetics, the "now-time" becomes sticky (2012); in Abdolah's work, the space of the present is an opaque, sticky cluster of ever-expanding associations with both past and present events, either remembered or fictional. For Glissant, who opposes the "right to opacity" to the Western quest for transparence in the name of appropriation and control, opacity stands for the irreducible density of the cultural other, a density which resists all efforts of appropriation (1997, 189–94). Opacity would result from the irreducible diversity of the world: "Transparency no longer seems like the bottom of the mirror in which Western humanity reflected the world in its own image. There is opacity now at the bottom of the mirror, a whole alluvium deposited by populations" (1997, 111). Abdolah also understands the past's presence in the present as an unrepresentable opacity. His sticky, intimate, even erotic relational aesthetics finds that meaning can only be produced by intimate interrelations with people who will, perhaps unwittingly, unveil part of a meaning they do not possess and which cannot be anticipated. Remarkably, it is the proverbial Dutch openness of women and gay men, such as René, that make this interaction possible.

Abdolah builds the story of his hero upon the Dutch characteristics of openness, though he refuses the realism and the notions of independent identity that come with it, and the related notion of tolerance that depends on a sense of hierarchy. Abdolah's semiotics does not lead to an insight into cultural difference. His story is about the phenomenon of difference itself (of being excluded, of being in-between) and, as he narrates from the very location of difference, its semiotics cannot be frustrated by Dutch absolutist definitions of difference as either presence or absence.

Ellen Ombre

Surinamese-Dutch writer Ellen Ombre also questions the assumption that the problems of integration will be solved if immigrants learn about Dutch culture, its history, and its values. Not unlike Abdolah's perception of knowledge as the product of interaction, Ombre argues that the multiplicity of

Dutch culture makes it impossible to know from the dominant perspective only.

In a collection of stories from 2000, *Valse verlangens* (False Desires), Ombre questions the possibility of knowing in a more fundamental way. She focuses on the Dutch perception of the easy, open verbal exchange as a means of sharing and producing knowledge—a perception only seemingly proximate to her own.

> They spoke as if they exhaled, effortlessly. We are made out of words, but I fall short. To converse is an art form. In that sense, I have not learnt how to speak, not about "the" this and "the" that. Not by birth (14).[14]

> The pleasure I had imagined taking from my education failed to materialize. Knowledge seemed to be restrained. One did nothing but talk. About one's own feelings and those of others . . . My supervisor, a bald creep with hairy hands, judged my language too posh (30).[15]

Openness is here unmasked as a specific rhetorical skill, in which one has to be trained. They are therefore excluded from public debates and the production of knowledge. The rhetorical tradition of openness is also highly evasive—within different institutions and social circles different modes of speech are demanded, or dismissed as not open enough, a practice that effectively silences and excludes those on the outside. But Dutch speech habits are also inadequate in addressing this reality. This is not just because the Dutch do not know their own colonial history, but also because they do not address the meaning of the hidden perversities in their present—perversities that maintain earlier colonial practices of abuse. The stories in this collection all address the present-day manifestations of the destructive colonial desires that have been analyzed by many, from Frantz Fanon, to Laura Ann Stoler, Robert Young, and Greg Thomas. In one story, a Dutch pedophile, a member of progressive multiculturalist circles, takes it completely for granted that his Surinamese female acquaintance sympathizes with his sexual obsession for an Asian child. Elsewhere, such openness in matters of sexuality is linked to the Dutch lack of shame.

Thus, Ombre offers a devastating critique of the Dutch values of tolerance and openness, as rearticulated within its current permissive society. To the extent that inherent bias relating to race, colonization, and slavery continue to have an unacknowledged impact on the Dutch conception of a multicultural society, they reinforce exploitative power hierarchies and suppress critical counterdiscourses. Finally, the tenacious Dutch habit of reducing all sexual and relational conflicts to cultural difference makes it impossible to address these complex issues of power and exploitation. Ombre attacks this habit in very explicit ways:

> "Your problem [relational problems within an interracial marriage] should really be reduced to the difference in cultures." [. . .]
> "You are just saying something," Humphrey said. He felt duped by this false observation. [. . .]
> "Cultural difference can lead to serious relational problems."
> He reminded Humphrey of Marlon Brando in Apocalypse Now (126).[16]

Instead of assuming that the conflicts that trouble present-day Dutch society are rooted in cultural differences, Ombre focuses on political differences. Imposing a knowledge of Dutch cultural habits and values on immigrant communities will not bring a clear understanding of these political differences any closer.

Ombre's stories address these issues explicitly, while also experimenting with the stylistic means that best convey a certain sense of social multiplicity that cannot be easily resolved because of the potential conflicts that shape it. In some of Ombre's richest stories (for example, "Teveel tegelijk" or "Verbinding"), she evokes a layered, pluralistic image of a community of people whose lives intersect and intermingle for precise historical or social reasons, their aims and interests often lacking any common ground. Often, the meetings between these characters with conflicting social and sexual aims are characterized by obliviousness and miscomprehension. Thus, Ombre describes the small community of an immigrant family with a hapless, insecure daughter who wants to marry at all costs, within the context of a society of raw commodification. Or she suggests, in a few sparse descriptions and portraits, masterfully, the transnational, (post)colonial time-space of both Surinam and the Netherlands as a context for an understanding of the migratory dilemma (Hoving, 2006). Her work is meaningful not primarily because of its alternative imagination of space and time, but because it is a well-defined social and political critique, responding to a well-defined dominant discourse of openness.

Hafid Bouazza

Where will this literary unmasking of Dutch practices of tolerance and openness lead us? One possible answer lies within the sardonic, energetic, irresistibly irreverent work of Hafid Bouazza, with which we made our acquaintance in chapter 8. Bouazza's topics are (sexual) openness and the mechanisms of exclusion and inclusion. Quite in contrast to Ellen Ombre's distrust of desire and openness, he fondly imagines the erotic aspects of the loss of boundaries and differentiations. In his celebrated novels *Salomon* (2001) and *Paravion* (2003), mesmerizing descriptions picture the multicultural summer parks of the Dutch capital, Amsterdam, not as yumyum-forests to feast on, but as gardens of sexual delight. In the last chapter, we saw that these parks are brimming with sensuality and intercultural confusion. For Bouazza, sexuality is the domain in which intercultural differences are fought: "Mamette assured him that there would not be a wide cultural gap between them, as she understood him and his culture, but the nature of the insurmountable problem did not appear to be cultural, but sexual—in the widest sense of the word" (164).[17]

One could read the many references to flowers, plants, and animals as a critical reference to the explicit presence of sexuality in Dutch public space. But Bouazza doesn't present his plants and flowers as merely a symbol for Dutch-ness. Let me explain this with the help of the eglantine. This flower plays an important role in one of his stories about a difficult intercultural love affair, where a libertine Dutch girl tries to reeducate her Moroccon lover in the art of passionate love (Bouazza, 1996). The young man, who is the story's focalizer and narrator, finds himself impotent under the influence of the woman's eloquent zeal. However, the eglantine, which symbolizes what is—to the unfortunate male lover—an emasculating cultural code, need not unequivocally be read as the symbol of passionate love in *European* medieval courtly love poetry. Scholars have argued that such European poetic traditions are heavily inspired by the innovative art of islamic Spain, and the eglantine thus also refers to an Arabic understanding of passionate love (De Graef and Louwerse, 2001, 184; Menocal, 1987, 80). By filling the Dutch townscape to excess with the always hybrid, always multicultural, always mutating plant and animal life, under the sign of the eglantine, Bouazza pictures a landscape in which migrants can recognize not just traces of their own cultural heritage, but also the anxious cultural conflicts that are often worked through in the fields of love and sexuality. Whereas, for Abdolah, the fragmentation of identity is somewhat dangerous, Bouazza shows that this loss of self can also lead to an intense eroticism,[18] which, in turn, encourages semiosis. Bouazza's language reveals the sexual pleasure that lurks within the anxiety of fragmentation and dissolution. In his own writing, this sexual pleasure can be indistinguishable from the poetic pleasure to be derived from his intense, sensual play with language. For Bouazza, sexual difference rather than cultural difference lies at the heart of multiculturalization. If the Dutch preoccupation with difference (as absolute, as essential) testifies to a well-researched anxiety about national dissolution, these authors suggest a different point of view: immigrants and natives share both an anxiety about, an informed critical distrust of, and a desire for openness.

This desired openness has nothing to do with a tolerance that is founded on an ingrained sense of superiority—that is a notion these writers unmask, reject, or declare irrelevant. Nor do these authors celebrate the exclusionary rhetorical traditions that go by the name of openness. Their desire for openness cannot be answered by merely acknowledging an approach to diversity that reifies cultural difference. They do ask, however, for the acknowledgement of the mechanisms of exclusions at work on many levels and within many domains within Dutch society.

CULTURE IS NOT THE ANSWER

These writers, instead of being exclusively engaged in the creation of an alternative imagining of multiplicity and opacity, can also be considered cultural activists. As such, they demand an acknowledgment of the mechanisms of exclusion at work on many levels and within many domains within Dutch society, and of the history of that exclusion that shaped that society, including its own narratives about its natural and cultural identity. Responding to that demand, however, is no easy task.

Tolerance may be regarded as the central value of Dutch national culture but, as it is a strategy of institutionalization rather than a consistent value, it does not lead to a shared practice of openness to difference, which would stimulate multiculturalization and creolization. The prohibition on discrimination, imposed by law, receives a different form within every specific domain and institution, with its own "fine-meshed constructions of symbolic boundaries, [and, IH] processes of in- and exclusion" (as van Ginkel, 1999, 306, has it). The scope of a national policy is simply too general to effectuate the desired changes involving integration and diversity in all those cultural contexts, which do not comprise one homogeneous national culture (van Ginkel, 306).[19]

When we look at the strategies adopted by young Dutch immigrants, as did three colleagues, including myself, in a research project on immigrant influences on Dutch everyday culture, we see that they do not merely try to find a place in society by emphasizing their ethnic cultural identity (Hoving, Dibbits, and Schrover, 2005). In this, they resemble the immigrant writers I have discussed. In reaction to an identitarian discourse that tries to categorize and situate young immigrants in ways they find confining, they take recourse in a variety of strategies of identification.[20] It is tempting to see this, also, as an effort at creating opacity—by refusing to self-identify in terms of only one or two dominant axes of categorization (nation, ethnicity).

Whether these literary and everyday counterstrategies of identification, which evade and critique the regulating mechanisms of cultural tolerance and integration, will be acknowledged and result in a strong counterdiscourse, depends on whether these voices will be heard in public and political debates. In this it seems that the oft-criticized neighboring nation of Germany is in advance of the Netherlands; immigrants are represented to a much larger extent in trade unions and work councils. The sense of being represented within the domain of politics, and therefore of enjoying "a shared engagement with the political society," may be a more important factor for one's sense of being included in society than shared cultural values, as Bhikhu Parekh argues in a Dutch immigrant journal (2003, 10). However, after the first decade of the twenty-first century, which saw the rise of Islamophobia and other forms of racism, exclusion, and xenophobia in the Netherlands, the gap between the different ethnic groups in the Netherlands has only been widened. If more immigrant voices are heard in different media, the voice of neorealist intolerance seems to drown most other, more modest voices in the public arena.

Within the Netherlands, the notion of tolerance implies that the acknowledgment of cultural rights is the main key to integration. But in a nation with a tenacious tradition of reducing power differences to "cultural" differences, and a tradition of regulating "cultural" difference into a segregated structure with its own unspoken hierarchies, neither the celebration of cultural identity nor the plea for multiculturalization based on a reified sense of identity will bring about change toward a genuinely plural society. What is needed is a subtle understanding of the vexed, interdisciplinary, paradoxical issue of representation, that opposes the populist, conservative neorealism that now marks public debates. Such an understanding implies, first, the insight that immigrants should have adequate social and political representation in the dominant institutions; second, the insight that the Dutch dominant discourses, organized around the notions of openness and tolerance, do not offer immigrants much space for cultural representation; third, the insight that the notion of representation needs to be, and is, problematized to the point of embracing notions such as opacity.

The fact that writers address the natural environment to criticize exclusionary *social* discourses shows that the mechanisms at work in social debate are part of a deep-seated cultural imagination. This dominant imagination of the Dutch nation as open and tolerant naturalized itself by appealing to its specific environment; but the dualist, essentialist dominant Dutch environmentalist discourse is, in turn, marked by the nation's political history of essentializing difference. When writers and critics problematize the concepts of openness and tolerance, they may move away from representation to anti-representational and anti-identitarian explorations of multiplicities, opacities, and flows (such as in Bouazza's stories of incessant transformation, or in Abdolah's refusal of realism in the representation of migration). However, this chapter has shown that these literary moves can best be understood as interventions in exclusionary social Dutch *discourses*. When these writers translate social and political differences into sexual differences, intercultural misunderstandings into an erotic semiotics, and sexuality into vegetable energy and the dissolution of self, they do not position themselves outside the realm of politics. To the contrary. They show that politics cannot be understood without addressing the natural environment too—as the site of the experience of dislocation, as the sensual space for disorientation and reorientation, as the foundation for discourses of exclusion and inclusion. To counter the Dutch perception of nature as absolute, and the foundation for an essentialized national identity, these writers do not only protest identitarianism and realism; they show that the Dutch environment is intensely historical, shaped by intense interaction, and still being shaped, imagined, and reimagined by the very real sensual, social, and political experiences of many very different people.[21]

NOTES

1. My translation of: Ik moest wennen. [. . .] Ook moest ik aan die overheersende kleur groen wennen. Een koe in de mist was nieuw voor mij. Aan die blote benen, buiken, borsten, billen en aan de taal moest ik wennen.

2. During the economic boom in the Netherlands at the end of the 1990s, the unemployment of immigrants was still three times as high as that of the autochthonous population, whereas in the first decade of the twenty-first century, during an economic crisis in Germany, their unemployment there has been only twice as high as that of the autochthonous population (Koopmans, 2003). Around the turn of the century, 35 percent of the Turkish pupils in the Netherlands left secondary school without a certificate, compared to 21–23 percent in Germany (boys-girls).

3. In the case of "gedogen," the acknowledgment is arbitrary; offenders can be prosecuted whenever the law finds it necessary to do so, without having to motivate the prosecution of the one and not the other. "Gedogen" leads to inequality before the law.

4. The impulse to institutionalize difference is one way to deal with a variety of differences, which are, reductively, interpreted as absolute differences in culture. This approach is, in fact, a form of passive tolerance. Conversely, the tendency to deny difference, to refuse to deal with it and to look the other way, is a passive form of intolerance. Seeing difference as merely a form of deprivation or backwardness can also be understood as a form of evasion (see Koenis, 2002, 74).

5. Some immigrants also felt discriminated against and insulted by the project ("I am not an ape," a Cape Verdian politician said).

6. One of the already all too obvious stars in such narratives would be the tulip, the flower that symbolized traditional Dutch national culture, but that can also be recognized as an apt symbol for the *hybrid* Dutch nation. Being a migrant from Turkey, transported for economic reasons, it refers to the many migrant workers from that region, whose arrival in the Netherlands heralded the intense multiculturalization of the Dutch nation in the 1960s. In this sense, the tulip fields offer wonderful multicultural narratives.

7. All translations of the Dutch literary texts are mine.

8. Ik? Ik ging om alles te zien. Het proces te volgen. Om erachter te komen hoe het zou kunnen verlopen als je verbannen bent.

9. Toen ik zag dat René zich zijn verleden niet goed meer herinnerde en dat hij niet meer wist dat ik Bolfazl was, besefte ik ineens dat ik ook die geur niet meer kon terugroepen.

10. Ik moest goed kijken, goed luisteren en alles onthouden. Ik zou ooit kunnen vertellen wat verbannen zijn betekent.

11. … ik zoek een overeenkomst tussen gebeurtenissen [. . .] Eerlijk gezegd vertrouw ik mijn herinneringen niet meer zo. Ik verzin al die verhalen zelf.

12. Fantasie en werkelijkheid lopen door elkaar bij een banneling.

13. Verhalen hebben hun eigen wetten. Men is niet in staat om de loop van de verhalen te veranderen.

14. Zij praatten alsof ze lucht uitbliezen, moeiteloos. We hangen van woorden aan elkaar, maar ik schiet tekort. Converseren is een kunst. Ik heb in die zin niet leren praten, niet over "de" dit en "de" dat. Van huis uit niet.

15. Het plezier dat ik me van de opleiding had voorgesteld, bleef uit. Kennis leek aan banden te worden gelegd. Men deed daar niets anders dan praten. Over eigen en elkaars gevoelens . . . Mijn scriptiebegeleider, een kale creep met haar op zijn handen, vond mijn taal te hoogdravend.

16. "Jullie probleem moet werkelijk worden teruggebracht tot het cultuurverschil!" "U zegt maar wat," zei Humphrey. [. . .] Hij voelde zich de dupe van die valse waarneming. [. . .]

"Cultuurverschillen kunnen tot ernstige relationele problemen leiden."

Hij deed Humphrey denken aan Marlon Brando in *Apocalypse Now*.

17. Mamette stelde hem gerust dat er geen culturele kloof tussen hen zou gapen, omdat zij hem en zijn cultuur begreep, maar het onoverkomelijke probleem bleek niet cultureel van aard, maar seksueel, in de ruimste zin van het woord.

18. See also chapter 3, *Aelred's Sin*.

19. One might wonder how the "poldermodel," with its emphasis on negotiation instead of open conflict, relates to the Dutch *inability* to really work through differences, through open, respectful negotiation. The answer lies in the nature of the "polder"-negotiation: it refers to the search for compromises between conflicting interests between different parties in long and extensively regulated (some say: inefficient) procedures. The Dutch government and the political powers of the day, however, still decide, through subtle forms of accommodation, "co-optation, and marginalization," which parties and minorities will be admitted as permanent partners in the procedures (De Haan, 2002, 80–90). The poldermodel does not concern the careful working through of ideological differences.

20. Some may define themselves variously as town dwellers (Amsterdammer instead of Dutch), as Muslim, as professional; some may lie about their ethnic identity, sometimes by way of play, sometimes in order to escape discrimination; some may refuse to define their mixed descent. They will often identify with professional or political role models (see also Prins, 2004), or with a certain music scène.

21. This chapter is based on two earlier essays, of which one was first published electronically as a Transit publication, based on a lecture at the University of California, Berkeley. http://german.berkeley.edu/transit/2005/TRANSIT50909.pdf.

Chapter Ten

Matter

The Last Word

This book was written to explain why, and how, a range of Caribbean writers read the postcolonial Caribbean and globalization through the lens of nature, the environment, earth, or surroundings. The fact that the material environment is central to their work already suggests that postcolonial and Caribbean sociological theories alone, and the theories of globalization that are inspired by postcolonialism, do not suffice to come to grips with the global condition of the world. In the preceding chapters, I have argued that the literary texts under scrutiny are often specifically concerned with the desire for the "real"; they are energized by the urge to relate to the global world's more-than-human materiality. This last chapter is meant to bring into focus the debate on this desire that runs through many of the above chapters. I will revisit the question regarding the theoretical framework needed to create a truly productive analysis of the evocation of nature's materiality in Caribbean writing.

We discussed three different literary means through which Caribbean texts evoke the environment in response to (neo)colonial violence: by deconstructing exclusionary and oppressive discourses on nature and the natural and developing counterdiscourses; by imagining the (human and non-human) world as intensely interconnected; and by evoking the shocking, extradiscursive, unrepresentable materiality of the earth. This last chapter will specifically discuss the pitfalls in theorizing the last two strategies: the imagination of interconnectedness, and the evocation of the material "real."

Traditionally, metropolitan ecocritics tended to look for an ethics of interconnectedness in environmental writing. An ethics that should reestablish human closeness to nature, to the world's "real." We have seen that postcolonial, and especially Caribbean, writers often found such an ethics of interconnectedness counterproductive. This study suggests that postcolonial and/or Caribbean literary texts rather aim at *coming to terms* with incurable dividedness, not so much because they want to overcome alienation, but because they want to explore trauma, and, equally important, address and resist oppressive patterns. This insight leads me to the conclusion that the concern with materiality is much more complex than the desire for reconnection. It can only be understood through a variety of theoretical approaches, that is, apart from (new) materialism, predominantly postcolonial theory, psychoanalysis, phenomenology, (trans)gender studies, and Caribbean and African-American epistemologies. In addition, a more sociologically and/or politically inspired approach appears indispensable to analyze the relation between these theoretical practices, and specific environmentalist action.

Let me return once more to the way in which, in literature, a (new) materialist reading can be put into dialogue with a psychoanalytical reading, phenomenology, and postcolonial criticism. In a famous essay on thing theory, literary theorist Bill Brown discussed the *fleeting*, unstable relation persons have with things (and Kincaid's experience of the "thingish" cotton flower described in chapter 5 is a good example) (Brown 2001). In spite of this unstable relation, which questions the status of things as autonomous, stable entities, we can very well adopt a "methodological fetishism," Brown proposes, inspired by Arjun Appadurai, and nevertheless study the effects that objects have on human beings, disregarding the work that has produced these objects' materiality in the first place. The effects objects have can as often be unsettling as stabilizing, as Kincaid's encounter with the cotton flower shows. Here, Brown comes close to new materialism. The nonhuman world affects us. However, as Brown reminds us, society also shapes the way in which we understand nature (and "matter," "things," etc.). Society socializes our sensual perception, and our "corporeal imagination" (Castoriades's term, in Brown, 9). But the corporeal imagination may also resist that socialization, perhaps by imagining the force of matter or nature *otherwise* (9–10). One of the ways in which that may be done is by vigorously evoking the sheer disruptive force of nonhuman matter—as Kincaid did, by dramatically opposing a flower's perfection to its implication in brutal history.

A remarkable observation in Brown's argument is his insight that, instead of indicating stable truth, things point at "a certain limit or liminality, to hover over the threshold between the nameable and the unnameable" (5). Though the above reading (chapter 5) of Kincaid's appreciation of the thingness of what she did not recognize as a cotton flower, in which phenomenology and materialism came together with the critique of colonial discourse, makes sense, we have also seen that this liminality can take on another, intense psychological significance. In the introduction, I discussed how the sudden experience of the Real often comes from trauma, just as the desire for a stable reality can be a response to social crisis. In chapter 7, I quoted Elizabeth Grosz's explanation of the differences between a psychoanalytical approach and a (new) materialist understanding of the relation between the human and the nonhuman—the first would assume a gap, the second continuity. Most of the novels I discussed operate in the liminal space, in the gap between what can be represented and what cannot. They often present a painful, traumatic gap that can sometimes only be healed by a return to matter, or death.

This observation suggests that the focus on harmonious interconnection, sometimes found in metropolitan ecocriticism, is inadequate. The mere acknowledgment of the interactive relation between the human and the nonhuman can hardly be a decisive beneficial alternative to the dualist discourses that caused and accompanied traumatic violence, as the advocates of a harmonious pastoral perspective might suggest. The devastating critiques of the notion of interconnectedness discussed in chapters 3 and 4 explain why not, and Nancy Tuana's and Stacy Alaimo's illuminating analyses of the implications of the human "viscous porosity" in a toxic environment are other points in case (Alaimo, 2010; Tuana, 2008). Their work shows that, while (new) materialist approaches may resist the psychoanalytical notions of the gap between humans and the rest of the world, they can also help to analyze the often painful and destructive, complex ways in which the world is both connected and disconnected.

The analyses in this study have shown that Caribbean literary writing has clear affinities with a new materialist approach. Some of the writers considered in this study, for example, would agree that the wish to revive the contact with nature's materiality is hard to realize on the level of discourse. Olive Senior and Jamaica Kincaid propose another way to relate to nature that, at first sight, seems to come close to the (new) materialist approach: not through representation, but through agency. In their approach, landscape becomes taskscape. This productive approach responds to the basic postcolonial *and* ecocritical aim to contribute to social change. Their cultural critique hopes to offer an ethics (Appiah) that might inspire action. Here, however, one of the crucial differences with new materialism comes to light. Especially Senior's work (just as that of Ellen Ombre) offers a marxist-inspired political reflection, rather than a philosophical one.

This is only one of the differences that catch the eye. The materialist approach, of which new materialism and Deleuze and Guattari's work are the most influential examples, strongly resonates with postcolonial, Caribbean, and African-American critical discourses—not in the least because some of their most important spokespersons have appropriated Deleuze's and Guattari's work (Glissant, Benitez-Rojo, Donatien-Yssa). Postcolonial, Caribbean, and African-American theories are in themselves syncretic, as they often also embody a working through of metropolitan thought. However, this is the very reason why they also represent a *critique* of the approaches they have appropriated.

In the following section, I will discuss some aspects of the very important, influential work of Edouard Glissant in relation to the ecological and environmental discourses with which his work has much in common. This will help to understand the problems that may emerge when (new) materialist thought is applied to analyze the Caribbean (political, economic, and environmental) condition. Above, I have often pleaded for the inclusion of psychoanalysis in the analysis of Caribbean writing. Here, I will consider the connection with social and political analysis. In addition, this discussion of Glissant's work will allow me to go back to the question of to what extent the medium of literature is appropriate for the exploration of the materialist turn.

POETICS, ETHICS, ACTION

Glissant's influential poetics of Relation sees all beings as participating in what Glissant calls the "tout-monde," the world-totality, or chaos-world. As we have seen, the concept of chaos relates Glissant to a large and varied group of scholars who are more or less loosely inspired by the same concept: scientists as well as philosophers, biologists, and environmentalists as well as sociologists. There is great variety between their theoretical approaches. If we consider a range of approaches in ecology and environmental philosophy that are organized around key concepts such as dynamism, fragmentation, randomness, individualism, and nonreductiveness, we find a definite emphasis on specificity and concreteness.[1] These scholars do not just criticize reductive generalizations and universality, but they also insist on the concrete individuality and untranslatable specificity of each environmental situation. Here we find a marked affinity with Glissant's early insistence on the opacity and untranslatability of every location.

At the same time, Glissant's work (and that of other scholars of globalization) also aims at imagining a totality. To understand what is at stake, we might approach this contradiction by considering the affinity of Glissant's work to earlier environmental discourses. This closeness is apparent in his use of terms like 'totality,' a certain form of holism, and his insistence on the possibility to know this totality intuitively, from the inside. The "tout-monde" doesn't ask for analysis but for awareness: "In spite of ourselves, a sort of 'consciousness of consciousness' opens us up and turns each of us into a disconcerted actor in the poetics of Relation" (1997, 27).

It may be useful here to remind us of the historical development in environmentalist philosophy.[2] The most ambitious environmental movement of the 1970s—the period in which Glissant had already established himself as a writer, poet, and critic, though his influential essay *Le Discours Antillais* would only be published in 1981—advocates a philosophical holism that emphatically strives toward an understanding of the world as a whole. This "deep ecology," a movement associated with Arne Naess, insists on seeing the world as a "relational, total-field image" (Naess, qtd. in Kirkman, 2002, 21), "a web of interconnections that has its own internal unity, its own goals and interests" (Kirkman, 8). Just as many organicist and holistic environmental philosophies, Glissant's poetics of Relation responds often explicitly (and critically) to Cartesian mechanicist discourse, and is close to the critiques of mechanicism and dualism of the last three decades.[3] In Europe and the United States, the new organicist approach was elaborated in theories which saw nature as a self-regulating, efficient system: climax theory and its successor systems ecology (Clements and Odum). This became the predominant view in the 1970s (Schouten, 2001; Keulartz, 1998; Marshall, 2002). It is a holistic, anti-mechanicist, organicist approach, though it did not break with dualism. In its most radical form, man is seen as exterior to nature, which is best left alone to develop to its greatest diversity. In Germany and in the United States, however, the philosophical holism celebrated by "deep ecology" lost much of its credibility because it appeared closely related to an undesired view of society: the submission of the individual to the state. Its very non-anthropocentrism attracted critique from socially committed ecologists.[4]

Some of the comments from the opponents to deep ecology are particularly enlightening. Let me refer to an example from the late twentieth century. In a 1993 essay, well-known philosopher, scientist, ecologist, and activist Vandana Shiva argues why she is deeply distrustful of unifying, totalizing discourses of the environment. Whereas in the 1970s and 1980s grass-root movements addressed local environmental crises, such as deforestation, which were identified as caused by multinational corporations and institutes, in the 1990s these concerns were reformulated as global environmental problems. A new agenda of global environmental problems appeared, reducing the many local environmental crises to only four (in the beginning of the 1990s). The destruction of the ozone layer was one of these. The term "global" conceals the real causes of this problem (which in this case would be the production of CFCs by a small group of transnational industries). By redefining the problem as global, it now becomes the problem and responsibility of the millions of consumers of environmentally destructive fridges and air-conditioning units—many of whom live in the South. What is more, the producers themselves, who are the cause of the problem, are now called upon to provide a solution—that is, technological alternatives to the CFCs (Shiva, 1993). Deep ecology does not seem to have much to offer environmental activists fighting this kind of global holism.

Such interventions invite us to think about the political agendas implicit in theoretical practices, as Galloway (2011) encourages us to do (introduction). Glissant's complex notion of relationality, which is primarily based on the work of Deleuze and Guattari, differs from the holism proposed by deep ecology in important aspects. Nevertheless, it is important to remain critical of the way in which a relational poetics borrows from deep ecology's philosophy, especially when both inscribe themselves in the phenomenological tradition. Many deep ecologists argue that the experience of the world's organic interconnectedness is intuitive, and can be sensed "from within"; it is produced by "a perceptual immersion in the surrounding world" (Abrams, qtd. in Kirkman, 2002, 80). One recognizes the phenomenological, and specifically Heideggerian, insight that this embodied experience would lead to an authentic way of being in the world, which could then serve as the foundation of an ethics (Kirkman, 140). This approach is speculative, not empirical, and it can therefore not be refuted through empirical and scientific means. It is, however, possible to consider its political agenda, as Galloway (2011) suggests, and it can then be said that it is far removed from the strategies of a political ecologist such as Shiva, and notably different from the much more critical view of totalizing discourses that can be found in the work of many other political ecologists.

Phenomenological insights can nevertheless help to articulate useful critiques. Just as Glissant, deep ecologist David Abram (1996) sees an intimate relation between the experience of nature as a "densely interconnected organic network" and language: "this dynamic, interconnected reality [lends] something of its structure to all our various languages" (qtd. in Kirkman, 2002, 80). Glissant, too, uses the analogy with language (notably creole language) to develop his theory of Relation, and he sees language—"la parole"—as the privileged way to inscribe oneself into the materiality of the world (1990, 202). In Glissant's view, there is no gap between the landscape, or the world, and the text: "Then words, no one's fiefdom, meet up with the materiality of the world. Relation is spoken" (202). As Jean-Pol Madou (1996) suggests, Glissant's text creates an almost tactile continuity between eye and world (75). Text and world are moved by the same dynamic, chaotic forces of Relation, and they are therefore linked metonymically.

This emphasis on the materiality of language and poetry as the privileged means to "be in the world" represents an important step away from the dualist patterns through which nature and matter are often thought as opposed to culture and discourse. It may even offer a welcome alternative to the focus on language and textuality (as opposed to materiality) that occurs in much postmodernist scholarship, as Alaimo and Hekman (2008) argue, along with many others in the field of environmental humanities (2–3). This view on language can also be read as a radical refusal of the contemporary all-invasive discourse of economy-driven technology, which informs the discourse of global environmentalism criticized by Shiva (1993) and others (Ziarek, 2002). Only poetry would be able to evade the technological and economic endeavors to incorporate all possible alternative discourses on the environment, and erase their difference (Kirkman, 2002, 140). This technological incorporation can be seen as a strategy of translation, as political ecologist and activist Larry Lohman (1993) has it: "[g]reen globalism appeals strongly to many Northern environmentalists [. . .] because it tries to translate all important 'environmental' practices and insights into a common, comfortably modern vocabulary. This globalism [. . .] satisfies a deep-felt Western [. . .] need for containment and control" (159). The defense of the value of specific, untranslatable languages might counter that strategy of control. Glissant's insistence on the *opacity* of every specific location and moment in the world, and of its untranslatability, seems to spring from the same insight. This shared resistance against a common vocabulary seems a more productive position than the celebration of creolization as an autonomous force which can also be found in Glissant, but which does little to oppose the universalism of globalism. The global market is perfectly at ease with notions of creolization.

This concise comparison between deep ecologists, political ecologists, and aspects of Glissant's poetics of Relation shows an ambivalence in Glissant's important project that points at a broader, more general problem in theorizing the global (environment). This ambivalence can be described as the tension between the emphasis on specificity, opacity, and untranslatability on the one hand, and the emphasis on the universal, which can both be found in Glissant.[5] The problem can also be located in the ambivalent status of the knowledge produced in Glissant's work, and this is what I aim to discuss here. If political ecologists suggest that *specific* (instead of abstract, general) knowledge of social and political structures and practices is necessary to counter global discourses and take action, Glissant criticizes their assumption that such knowledge would be useful—though, on the other hand, he *does* claim to produce knowledge, stating that he wants to reconnect "with poetry's earlier ambition to establish itself as knowledge" (Glissant, 1990, 138). However, he takes issue with the (deep ecologist's) claim that this knowledge might serve as the foundation for an ethics. He emphasizes all the gaps in the chain poetry-knowledge-ethics-acting, without diminishing the importance of each element: "Let us not stop with this commonplace: that a poetics cannot guarantee us a concrete means of action. But a poetics, perhaps, does allow us to understand better our action in the world" (199). As Natalie Melas (2007) has emphasized, Glissant offers a poetics, and not an ethics of Relation. This in itself is clear. The confusion comes from his readers' understandable assumption that the knowledge produced by this poetics would be the kind of knowledge that would be useful from the perspective of Glissant's own long-standing political engagement with the marginalized and the excluded. Yet the kind of knowledge offered by the later Glissant appears to be of a much more abstract and general kind. As Hallward (2001) has it, "the *fact* of dispossession itself—has not changed. [. . .]

Glissant's critique of dispossession risks conversion into an effective affirmation of dispossession (however 'positively' affected)" (120). Glissant's readers should perhaps not seek for more than a poetics has to offer.

It is instructive to compare Glissant's work to contemporary ecocritical efforts to theorize globalization, or the contemporary global awareness. Ursula Heise's well-known 2008 study on the imaginations of the planet is a good example. It adopts many of the points of departure we also find in Glissant, and in many other postmodern ecocritics: it emphasizes the need to steer free from holism and illusions of stability, and to acknowledge dynamism, deterritorialization, crisis, and interactivity. At the same time, however, it insists on the need to consider the *actual* experiences of living in the globalizing world. Heise's study testifies to the widely shared acknowledgment that a theory of globalization (including its environmental aspects) can only have explanatory force when it is concerned with the specific. When it is, it results in the most productive moments in her own *and* Glissant's work.

I hope, in my discussion, to have highlighted some of the difficulties in establishing a link between the articulation of a poetics or a theory of globalization, an ethics, and a politically successful form of agency. Yet ecocritical scholarship still insists that the imagination, or thought, is the privileged point of departure: "the political will to implement new greener technologies will only come [. . .] from a popular shift in ways of thinking about our relationship with our home planet," as the hopeful editors of a 2007 collection of ecocritical writing suggest (Becket and Gifford, 7). It seems to me, however, that, even if literary and visual imaginations can successfully intervene in public discourses on the environment, political change will only come about by tenacious, varied, collective political action. The complex relations between a certain "way of thinking" on the one hand, and the political, economic, and environmental organization of society on the other, have been the topic of intense scholarly debate for more than a century, and one of the provisionary conclusions seems to be that there is no direct link between "beautiful thoughts" and good, effective action. On a smaller scale: MacNaghten and Urry (1998) offer wonderful examples of the complex social and political settings in which environmental change does actually occur (219–46).

WHAT WRITING CAN DO

This brings us again to the question of what literature *does* have to offer to the world. To answer that question, I would like to go back to the desire for the "real" that lies at the interface of creative writing and the nondiscursive world. Above, we related that desire to trauma. Here, I want to consider the connection with the writing process itself. In an essay on the representation of things in literature, literary scholar Peter Schwenger (2001) states: "In the representation there always remains a drive toward something beyond what is represented, something more than mimesis" (109). He then equates this drive with the death drive, which also represents something (in ourselves) that cannot be grasped, and that cannot be satisfied (and cannot be represented). "So Lacan identifies the death drive with a shortfall in representation" (109). One might object that this drive could also be aimed at pleasure, and that remark does indeed come up in Schwenger's analysis of one of Francis Ponge's poems. He answers that objection by quoting Ponge's own remark about the writing process, which "must be thought of: not as the transcription, according to conventional rules, of some idea [. . .] but, in reality, as an orgasm: as the orgasm of being or structure, let's say" (Ponge, qtd. in Schwenger). "Such a fulfillment is also an annihilation," comments Schwenger; "a *petite mort*" (110). Indeed, just like the desperate desire to reestablish one's contact with nature, with the real, stable matter of life, the death drive comes down to the wish to return to matter and escape the tortuous uncertainties and tensions of one's multiply mediated, torn existence. This drive can well be fuelled by trauma, as has been my point throughout the book. But it can also be connected to the general condition of the postcolonial, globalizing world. As Jamaica Kincaid (1999) puts it, "the world is cracked, unwhole, not pure, accidental" (*Garden*, 92). There is no escape. And there is no end to the desperate work with this desire, in writing, either.

This endless manipulation of libidinal energies is one of the layers, or dimensions, of creative writing. The presence of this dimension is one of the reasons why literature is so often much more wayward, vicious, and complex than if it were only interested in offering an ethical example. This libidinal economy, to use Lyotard's words, may even confuse a text's work with images of nature (perhaps meant to critique or undo certain discourses) to the extent that it explodes its reliance on discourse or narrative altogether (see van Alphen [1992] on the way in which painter Francis Bacon resists narrativity). An ethical approach to literature (which may be found in some phenomenologically inspired ecocriticism) could result in blindness to writing's negative, destructive energies. A new materialist approach, however, which would *only* take into account the text's energetic flow, risks to overlook its discursive positioning—for even if the text's libidinal force seems to overrule its discursive significance, that referential element will still be there; no work with language can escape its intertextual significance. Therefore, a text that seems subversive because of its radical investment in the play with radical becoming may not be productive as a cultural critique when it is read on another level.[6]

Yet, the question to what extent literature can be seen as an alternative form of knowledge of the environment (in contrast to the knowledge yielded by dualist, reductive science) may not be so easily dismissed. By way of a last word, I will suggest an answer by first reconsidering the question whether man and nature are irretrievably divided by a gap, or whether they live in intimate continuity. Psychoanalysts will emphasize the divide, phenomenologists and new materialists will point out the continuity. I will make my point with the help of a Dutch theoretical intervention. Dutch evolutionary specialist Chris Buskes builds on the work of zoologist Lorenz to defend a realist standpoint. Though Buskes regards the apriori categories of space, time, and causality as mere working hypotheses, which are only valid within a certain cognitive niche, he agrees with Lorenz that they allow human beings to perceive reality as it is.[7] Buskes then suggests that we could see the organism's adaptation to its environment as a form of knowledge itself. This knowledge is interactive. It translates and expresses a certain condition or process in another medium. For example, the presence of calcium in the soil is expressed by the color and form plants assume. Buskes' poetic description of knowledge comes remarkably close to new materialism: knowing *is* acting; a certain growth *is* a modality of knowing the soil in which it grows; this growth does not occur on the foundation of knowledge, but, as *relating* to the soil's relationality, it is itself knowledge. Thus, it seems that popular cartoons or films that stage speaking pets or trees quite miss the point. To put it simply, the tree's materiality *is* its knowledge.

This is the way in which we can also approach literature's specific way of knowing. Cathy Caruth's (1996) statement that literature "is interested in the complex relation between knowing and not knowing" (3) is of pivotal importance here. The unique contribution of literature lies, in the first place, in the fact that it works in the very space where representation and that which exceeds it, knowing and not-knowing, collide. On its libidinal level, it responds to the world's forces, just like any other organism (in this case, through the medium of language). It is here that the interplay between knowing and not-knowing is most striking. But the text combines this first response, this first form of knowledge, with a second form of knowledge, on a metalevel. It is at this performative level that the text creates a structure of address. It comments on and translates its own libidinal energies in a discursive form, shaping it in such a way that it responds to social discourse. It is here that literature may receive the shape of social comment, and act as an intervention in a society's imagination of nature, for example.

The way in which literature contributes to the project of creating a postcolonial, posthumanist understanding of the world is therefore not through representation alone. Indeed, creative writing may have the (libidinal) aim to explode destructive or irrelevant discourses, and escape narrativity or discursivity altogether. Many of the case studies in this book show that references to the world's raw materiality are a valued means to accomplish that aim. This delegitimization is also realized by the writers's insistence that globalization can best be known, or related to, by facing the specific, *material* violence it causes, in the environment and in human or animal bodies. This literary engagement with matter severely problematizes the discourses that legitimize traumatizing colonial and global violence. This is one of the most importance reasons why references to matter, the environment, nature, are indispensable for nonmetropolitan writers who address the colonial heritage of the global contemporary.

Literature seems exceptionally well-equipped to address the social, discursive dimension of the global, while simultaneously responding to the world through the manipulation of its libidinal energies. Though literature is still the focus of ecocriticism, and with good reason, we might finally want to ask the question whether other media are not equally, or perhaps even more successful in evoking the shifting relations to globalization. Deleuzian approaches are much more popular in film studies, for example, than in literary criticism (but see Gelas and Micolet), and the materialist scholars mentioned in this study often prefer to consider visual media, if they think art analysis at all relevant for their inquiries. The growing relevance of the new, often digital media for the way in which worldwide audiences imagine the human relation to the environment suggests that the questions what *kind* of knowledge art produces and what its relation to social action can be are in continuous need of rephrasing.

NOTES

1. Henry Allen Gleason's (1926) evolutionary ecology may count as an early example, individual proposals such as Alan Marshall's (2002) deconstructionist postmodern associationism as a late-twentieth-century case.
2. This section is based on overviews by De Jong, Keulartz, Kirkman, Marshall, and Schouten.
3. These were articulated in disciplines as varied as science, ecology, and philosophy, in the wake of more general scientific revolutions at the beginning of the century that signified a break with Newtonian mechanicist science (Einstein's theory of relativity, quantum physics, and big bang theory). In response, and through intense debates between ecologists and philosophers, organicist models were proposed, which assumed that living organisms, rather than being machines, were hierarchically organized wholes, which could not be reduced to the sum of their parts.
4. Socially committed ecologists form another tradition in radical ecology after the second world war, offering a completely different approach to the global system (the movement can be differentiated in social ecologists and political ecologists, such as Ivan Illich).
5. This is also a tension explored admirably by Natalie Melas (2007) and Peter Hallward (2001).
6. It will be clear that my approach to literature, in this book, is at odds with the approach proposed by E. O. Wilson in *Consilience*, who understands literature under the evolutionary denominator of "beauty." My analysis shows, I hope, that literature is most definitely *not* primarily concerned with beauty.
7. Outside that niche (as in, for instance, the field of quantum mechanics) they lose their validity (Buskes, 2006, 251).

Bibliography

Abdolah, Kader. 1997. *De reis van de lege flessen* (The Journey of the Empty Bottles). Breda: De Geus.

Abram, David. 1996. *The Spell of the Sensuous: Perception and Language in a More-Than-Human World.* New York: Vintage.

Achterhuis, Hans. 1993. "Scarcity and Sustainability." In Wolfgang Sachs (ed.), *Global Ecology: A New Arena of Political Conflict*, 104–16. London: Zed Books.

Adams, Rachel, and David Savran, eds. 2002. *The Masculinity Studies Reader*, 188–200. Malden, MA: Wiley-Blackwell.

Agosto, Noraida. 1999. *Michelle Cliff's Novels: Piecing the Tapestry of Memory and History*, New York: Peter Lang.

Alaimo, Stacy. 2010. *Bodily Natures: Science, Environment and the Material Self.* Bloomington: Indiana University Press.

Alaimo, Stacy, and Susan Hekman, eds. 2008. *Material Feminisms*. Bloomington: Indiana University Press.

Alba, Richard, and Victor Nee. 2003. *Remaking the American Mainstream: Assimilation and Contemporary Immigration.* Cambridge, MA: Harvard University Press.

Alexander, Jacqui. 2006. *Pedagogies of Crossing: Meditations on Feminism, Sexual Politics, Memory, and the Sacred.* Durham, NC: Duke University Press.

Allewaert, Monique. 2013. *Ariel's Ecology: Plantations, Personhood, and Colonialism in the American Tropics*. Minneapolis: University of Minnesota Press.

Alphen, Ernst van. 1992. *Francis Bacon and the Loss of Self.* London: Reaktion Books.

———. 2000. *Armando: Shaping Memory*. Rotterdam: NAi Uitgevers.

Amatmoekrim, Karin. 2007. "Decaloog van een oude vrouw" ("An Old Woman's Ten Commandments"). In *Waarover we niet moeten praten*, 127–37. Haarlem: In de Knipscheer.

Appadurai, Arjun. 1996. *Modernity at Large: Cultural Dimensions of Globalization.* Minneapolis: University of Minnesota Press.

———. 2000. "Grassroots Globalization and the Research Imagination." *Public Culture 12*: 1–19.

Appiah, Kwame. 1991. "Is the Post- in Postmodernism the Post- in Postcolonial?" *Critical Inquiry 17*: 336–57.

Arendt, Hannah. 1963. *On Revolution*. New York: Viking Press.

Astell, Ann W. 2004. "Saintly Mimesis, Contagion, and Empathy in the Thought of René Girard, Edith Stein, and Simone Weil." *Shofar: An Interdisciplinary Journal of Jewish Studies 22*(2): 116–31.

Attridge, Derek. 2014. "Contemporary Afrikaans Fiction and English Translation: Singularity and the Question of Minor Languages." In Birgit M. Kaiser (ed.), *Singularity and Transnational Poetics*, 61–78. New York and London: Routledge.

Babich, Babette E. 2006. *Words in Blood, Like Flowers: Philosophy and Poetry, Music and Eros in Hölderlin, Nietzsche, and Heidegger*. Albany: State University of New York Press.

Bagemihl, Bruce. 1999. *Biological Exuberance: Animal Homosexuality and Natural Diversity*. New York: St. Martin's Press.

Bahtin, M. M. 1988. *The Dialogic Imagination*. Edited by Michael Holquist. Austin: University of Texas Press.

Bal, Mieke, and Miguel Á. Hernández-Navarro, eds. 2012. *Art and Visibility in Migratory Culture: Enacting Conflict and Resistance, Aesthetically*. Amsterdam, New York: Rodopi.

Bannett, Eve Tavor. 1997. "Analogy as Translation: Wittgenstein, Derrida, and the Law of Language." *New Literary History 28*(4): 655–72.

Barad, Karen. 2003. "Posthumanist Performativity: Toward an Understanding of How Matter Comes to Matter." *Signs: Journal of Women in Culture and Society 28*(3): 801–31.

———. 2007. *Meeting the Universe Halfway: Quantum Physics and the Entanglement of Matter and Meaning*. Durham, NC: Duke University Press, 2007.

Basch, Linda, Nina Glick Schiller, and Cristina Szanton Blanc. 1994. *Nations Unbound: Transnational Projects, Postcolonial Predicaments, and Deterritorialized Nation-States*. Amsterdam: OPA.

Bauer, George H. 1997. "Gay Incipit: Botanical Connections, Nosegays, and Bouquets." In Dominique D. Fisher and Lawrence R. Schehr (eds.), *Articulations of Difference*, 64–82. Stanford, CA: Stanford University Press.

Baumann, Gerd, and Andre Gringrich, eds. 2004. *Grammars of Identity/Alterity: A Structural Approach*. New York and London: Berghahn.

Beckett, Samuel, and Georges Duthuit. 1965. *Proust. Three Dialogues*. London: Calder and Bogars.

Beckett, Fiona, and Terry Gifford, eds. 2007. *Culture, Creativity, and Environment: New Environmentalist Criticism*. Amsterdam, New York: Rodopi.

Becquer, Marcos and José Guatti. 1991. "Elements of Vogue." *Third Text 16/17*: 65–81.

Bennington, Geoffrey. 1988. *Lyotard: Writing the Event*. Manchester: Manchester University Press.

Berger, James. 2005. "Falling Towers and Postmodern Wild Children: Oliver Sacks, Don DeLillo, and Turns against Language." *PMLA: Publications of the Modern Language Association of America 120*(2): 341–61.

Bergthaller, Hannes. 2006. "'Trees Are What Everyone Needs': *The Lorax*, Anthropocentrism, and the Problem of Mimesis." In Catrin Gersdorf and Sylvia Mayer (eds.), *Nature in Cultural and Literary Studies: Transatlantic Conversations on Ecocriticism*, 155–76. Amsterdam: Rodopi.

———. 2011. "Everything Is Connected to Everything Else . . . But Some Things Are More Connected Than Others: On the Ecocritical Relevancy of Niklas Luhmann's Theory of Social Systems." Paper presented at the Ninth Biennial Conference of the Association for the Study of Literature and the Environment, Bloomington, Indiana, June 21–26.

Bernard, Louise. 2002. "Countermemory and Return: Reclamation of the (Postmodern) Self in Jamaica Kincaid's *The Autobiography of My Mother* and *My Brother*." *Modern Fiction Studies 48*(1): 113–38.

Bersani, Leo. 1987. "Is the Rectum a Grave?" *October 43*: 197–222.

Bhabha, Homi. 1984. "Of Mimicry and Man: The Ambivalence of Colonial Discourse." *October 28*: 125–33.

———. 1994. *The Location of Culture*. London and New York: Routledge.

Blair, Andrea. 2002. "Landscape in Drag: The Paradox of Feminine Space in Susan Warner's The Wide, Wide World." In Steven Rosendale (ed.), *The Greening of Literary Scholarship: Literature, Theory, and the Environment*, 111–30. Iowa City: University of Iowa Press.

Blerck, Henk van, and Harry Harsema, eds. 2001. *Natuur is de klei van mijn boeken: Schrijvers en natuur*. Publication for a national exposition. Schokland: Staatsbosbeheer.

Böcker, Anita, and Dieter Tränhardt. 2003. "Einbürgerung und Mehrstaatigkeit in Deutschland und den Niederlanden." In D. Tränhardt and U. Hunger (eds.), "Migration im Spannungsfeld von Globaliserung und Nationalstaat," *Leviathan Sonderheft 22*/2003, 117–34. Wiesbaden, Germany: Westdeutscher Verlag.

Bodegraven, Robbert. 1998. "Het Nederlandse landschap is niet meer dan lucht: Migrantenschrijvers over boekenweek-thema Panorama Nederland." *Onze wereld 41*(3): 20–22.

Bolaffi, Guido, et al. 2003. *Dictionary of Race, Ethnicity and Culture*. London: SAGE.

Bongie, Chris. 1998. *Islands and Exiles: The Creole Identities of Post/Colonial Literature*. Stanford, CA: Stanford University Press.

Boone, Joseph. 2001. "Vacation Cruises; or, The Homoerotics of Orientalism." In John C. Hawley, *Postcolonial, Queer: Theoretical Intersections*, 43–78. Albany: State University of New York Press.

Botkin, D. B. 1990. *Discordant Harmonies*. New York: Oxford University Press.

Bouazza, Hafid. 1996. *De voeten van Abdullah*. Amsterdam: Arena.

———. 2001. *Salomon*. Amsterdam: Prometheus.

———. 2003. *Paravion*. Amsterdam: Prometheus.

Bow, Leslie. 2009. "Transracial/Transgender: Analogies of Difference in *Mai's America*." *Signs 35*(1): 75–103.

Brathwaite, Edward K. 1971. *The Development of Creole Society in Jamaica, 1770–1820*. Oxford: Clarendon Press.

———. 1984. *History of the Voice: The Development of Nation Language in Anglophone Caribbean Poetry*. London: New Beacon.

Braziel, Jana Evans. 2009. *Caribbean Genesis: Jamaica Kincaid and the Writing of New Worlds*. Albany: State University of New York Press.

Brennan, Timothy. 2003. "The Empire's New Clothes." *Critical Inquiry 29*(2): 337–67.

Bright, Chris. 1999. *Life Out of Bounds: Bio-Invasions in a Borderless World*. London: Earthscan Publications.

Brown, Bill. 2001. "Thing Theory." *Critical Inquiry 28*(1): 1–21.

Bryant, Levi, Nick Srnicek and Graham Harman, eds. 2011. *The Speculative Turn: Continental Materialism and Realism*. Melbourne: Re.press.

Buell, Lawrence. 2005. *The Future of Environmental Criticism: Environmental Crisis and Literary Imagination*. Chichester: Wiley-Blackwell.

Buell, Lawrence and Ursula K. Heise, and Karen Thornber. 2011. "Literature and Environment." *Annual Review of Environment and Resources 36*: 417–40.

Burke, Carolyn, and Jane Gallop. 1985. "Psychoanalysis and Feminism in France." In Hester Eisenstein and Alice Jardine (eds.), *The Future of Difference*, 106–21. New Brunswick, NJ: Rutgers University Press.

Buskes, Chris. 2006. *Evolutionair denken: De invloed van Darwin op ons wereldbeeld*. Amsterdam: Uitgeverij Nieuwezijds.

Caminero-Santangelo, Byron. 2014. *Different Shades of Green: African Literature, Environmental Justice, and Political Ecology*. Charlottesville: University of Virginia Press.

Caminero-Santangelo, Byron and Garth Myers, eds. 2011. *Environment at the Margins: Literary and Environmental Studies in Africa*. Athens: Ohio University Press.

Cándani. 2001. *Oude onbekenden*. (Old Unknowns). Haarlem: In de Knipscheer.

———. *Huis van as*. 2002. (House of Ashes). Haarlem: In de Knipscheer.

Caruth, Cathy. 1996. *Unclaimed Experience: Trauma, Narrative and History*. Baltimore, MD: John Hopkins University Press.

Casteel, Sarah Phillips. 2003. "New World Pastoral: The Caribbean Garden and Emplacement in Gisèle Pineau and Shani Mootoo." *Interventions 5*(1): 12–28.

Certeau, Michel de. 1984. *The Practice of Everyday Life*. Berkeley: University of California Press.

Césaire, Aimé. 1956. *Logboek van een terugkeer naar mijn geboorteland*. Trans. Simon Simonse. Haarlem: In de Knipscheer. Bilingual edition of *Cahier d'un retour au pays natal*. Paris: Ed. Présence Africaine.

Chakrabarty, Dipesh. 2012. "Postcolonial Studies and the Challenge of Climate Change." *New Literary History 43*: 1–18.

Chancy, Myriam J. A. 1993. "Exile and Resistance: Retelling History as a Revolutionary Act in the Writings of Michelle Cliff and Marie Chauvet." *Journal of Caribbean Studies* 9(3): 266–92.

Cheney, Jim. 1994. "Nature/Theory/Difference." In Karen J. Warren (ed.), *Ecological Feminism*, 158–78. London and New York: Routledge.

Childs, Peter. 2006. "Householders: Community, Violence and Resistance in Three Contemporary Women's Texts." In Gerry Smyth and Jo Croft (eds.), *Our House: The Representation of Domestic Space in Modern Culture*, 175–94. Amsterdam and New York: Rodopi.

Choenni, Chan E. S., and Kanta Sh. Adhin, eds. 2003. *Hindostanen: van Brits-Indische emigranten via Suriname tot burgers van Nederland* (Hindustani's: From British-Indian Emigrants via Surinam to Citizens of the Netherlands). Den Haag: Sampreshan.

Cilano, Cara, and Elizabeth DeLoughrey. 2007. Against Authenticity: Global Knowledges and Postcolonial Ecocriticism." *Interdisciplinary Studies in Literature and Environment 14*(1): 71–87.

Clark, Timothy. 2005. *The Poetics of Singularity: The Counter-Culturalist Turn in Heidegger, Derrida, Blanchot and the Later Gadamer*. Edinburgh: Edinburgh University Press.

———. 2011. *Cambridge Introduction to Literature and the Environment*. Cambridge: Cambridge University Press.

Cliff, Michelle. 1980. *Claiming an Identity They Taught Me to Despise*. Watertown, MA: Persephone Press.

———. 1984. *Abeng*. Trumansburg, NY: Crossing Press.

———. 1987. *No Telephone to Heaven*. New York: Dutton.

———. 1993. *Free Enterprise*. New York: Dutton.

———. 1994. "History as Fiction, Fiction as History." *Ploughshares 20*(2–3): 196–202.

———. 2009. *Everything Is Now*. Minneapolis: University of Minnesota Press.

Coetzee, J. M. 1988. *White Writing: On the Culture of Letters in South Africa*. New Haven, CT, and London: Yale University Press.

Colebrook, Claire. 2000. "From Radical Representations to Corporeal Becomings: The Feminist Philosophy of Lloyd, Grosz, and Gatens." *Hypatia 15*(2): 76–93.

———. 2001. *Gilles Deleuze*. New York, London: Routledge.

Collett, Anne. 2002. "A Snake in the Garden of *The New Yorker*? An Analysis of the Disruptive Function of Jamaica Kincaid's Gardening Column." In Gerhard Stilz (ed.), *Missions of Interdependence: A Literary Directory*, 95–106. Amsterdam and Atlanta: Rodopi.

Cosgrove, Denis E. 1984. *Social Formation and Symbolic Landscape*. London and Sydney: Croom Helm.

Courtman, Sandra, ed. 2004. *Beyond the Blood, the Beach, and the Banana: New Perspectives in Caribbean Studies*. Kingston and Miami: Ian Randle Publishers.

Covi, Giovanna. 2003. *Jamaica Kincaid's Prismatic Subjects: Making Sense of Being in the World*. London: Mango Publishing.

Craps, Stef, and Gert Buelens, eds. 2008. *Postcolonial Trauma Novels*. Denton: University of North Texas Press.

Cruz-Malavé, Arnaldo, and Martin F. Manalansan IV, eds. 2002. *Queer Globalizations: Citizenship and the Afterlife of Colonialism*. New York and London: New York University Press.

Cudjoe, Selwyn R., ed. 1990. "Jamaica Kincaid and the Modernist Project: An Interview." In Selwyn R. Cudjoe (ed.), *Caribbean Women Writers: Essays From the First International Conference*, 215–32. Wellesley, MA: Calaloux Publications.

Culler, Jonathan. 1994. "Introduction: What's the Point?" In Mieke Bal and Inge E. Boer (eds.), *The Point of Theory*, 13–19. Amsterdam: Amsterdam University Press.

Cuomo, Christine J. 1994. "Ecofeminism, Deep Ecology, and Human Population." In Karen J. Warren (ed.), *Ecological Feminism*, 88–105. London and New York: Routledge.

Danticat, Edwidge. 1998. *The Farming of Bones*. New York: Penguin Books.

Davies, Carole Boyce. 1994. *Black Women, Writing, and Identity: Migrations of the Subject*. Abingdon and New York: Routledge.

Davis, Diana K., and Edmund Burke III. 2011. *Environmental Imaginaries of the Middle East and North Africa*. Athens: Ohio University Press.

De Haan, Ido. 2002. "Tocqueville in Nederland: Verzuiling, democratie en sociale cohesie in de Nederlandse geschiedenis." In Teun Jaspers and Joyce Outshoorn (eds.), *De bindende werking van concepten: reflecties over participatie, binding en betrokkenheid in opvoeding en onderwijs, arbeid en zorg*, 49–96. Amsterdam: Aksant.

DeCaires Narain, Denise. 1996. "Olive Senior." In Bernth Lindfors (ed.), *Twentieth-Century Caribbean and Black African Writers*, 340–48. Detroit: Gale Press.

DeLoughrey, Elizabeth M. 2007. *Routes and Roots: Navigating Caribbean and Pacific Island Literatures*. Honolulu: University of Hawai'i Press.

DeLoughrey, Elizabeth M., Renée K. Gosson, and George B. Handley, eds. 2005. *Caribbean Literature and the Environment: Between Nature and Culture*. Charlottesville: University of Virginia Press.

DeLoughrey, Elizabeth M., and George B. Handley, eds. 2011. *Postcolonial Ecologies: Literatures of the Environment*. Oxford: Oxford University Press.

Dimock, Wai Chee. 2006. *Through Other Continents: American Literature through Deep Time*. Princeton, NJ: Princeton University Press.

Dolby, Nadine. 2003. "A Small Place: Jamaica Kincaid and a Methodology of Connection." *Qualitative Inquiry 9*(1): 57–73.

Donatien-Yssa, Patricia. 2007. *L'exorcisme de la blès: Vaincre la souffrance dans Autobiographie de ma mère de Jamaica Kincaid.* Paris: Ed Le Manuscrit.

Donnell, Alison. 1995. "She Ties Her Tongue: The Problems of Cultural Paralysis in Postcolonial Criticism." *ARIEL 26*(1): 101–16.

———. 2005. *Twentieth-Century Caribbean Literature: Critical Moments in Anglophone Literary History.* London and New York: Routledge.

Down, Lorna. 2007. "'Stepping out of the Kumbla': Kincaid's AIDS Narrative, 'My Brother.'" *Caribbean Quarterly 53*: 16–29.

Duyvené de Wit, Tom, and Ruud Koopmans. 2001. "Die politisch-kulturelle Integration ethnischer Minderheiten in den Niederlanden und Deutschland." *Forschungsjournal Neue Soziale Bewegungen 14*: 26–41.

Edlmair, Barbara. 1999. *Rewriting History: Alternative Versions of the Caribbean Past in Michelle Cliff, Rosario Ferré, Jamaica Kincaid, and Daniel Maximin*. Wien: Braumüller.

Edmond, Ron. 2000. "Home and Away: Degeneration in Imperialist and Modernist Discourse." In Howard J. Booth and Nigel Rigby (eds.), *Modernism and Empire*, 39–63. Manchester and New York: Manchester University Press.

Edmondson, Belinda. 1993. "Race, Privilege, and the Politics of (Re)Writing History: An Analysis of the Novels of Michelle Cliff." *Callaloo: A Journal of African American and African Arts and Letters 16*(1): 180–91.

Edwards, Brent Hayes. 2003. *The Practice of Diaspora: Literature, Translation, and the Rise of Black Internationalism*. Cambridge, MA: Harvard University Press.

Edwards, Justin D. 2007. *Understanding Jamaica Kincaid.* Columbia: University of South Carolina Press.

Eeden, Ed van, et al, eds. 2004. *Want je ziet niet alleen wat je ziet: Over het werk van Koos van Zomeren*. Amsterdam: Arbeiderspers, 2004.

Emery, Mary Lou. 2007. *Modernism, the Visual, and Caribbean Literature.* Cambridge: Cambridge University Press.

Fanon, Frantz. 1952. *Peau noire, masques blancs*. Paris: Editions du Seuil.

Feder, Helena. 2002. "Ecocriticism, New Historicism, Roman Apostrophe." In Steven Rosendale (ed.), *The Greening of Literary Scholarship: Literature, Theory, and the Environment*, 42–58. Iowa City: University of Iowa Press.

Franklin, Sarah, Lucie Lury, and Jackie Stacey. 2000. *Global Nature, Global Culture*. London: SAGE.

Fumagalli, Maria Cristina, Bénédicte Ledent, and Roberto del Valle Alcalá, eds. 2014. *The Cross-Dressed Caribbean: Sexual Politics after Binarism*. Charlottesville: University of Virginia Press.

Gaard, Greta. 1997. "Towards a Queer Ecofeminism." *Hypatia 12*(1): 114–37.

———. 2010. "New Directions for Ecofeminism: Toward a More Feminist Ecocritics." *Interdisciplinary Studies in Literature and Environment 17*(4): 643–65.

Galen, John Jansen van. 1995. *Kapotte plantage: Suriname een Hollandse erfenis* (Broken plantation: Surinam a Dutch Heritage). Amsterdam: Balans, 1995.
Galloway, Alexander R. 2011. "Quentin Meillassoux, or The Great Outdoors." *French Theory Today: An Introduction of Possible Futures*. Pamphlet 4. New York: TPSNY/Erudio Editions. http://cultureandcommunication.org/galloway/FTT/French-Theory-Today.pdf. Last accessed May 5, 2015.
Ganguly, Keya. 1992. "Migrant Identities: Personal Memory and the Construction of Selfhood." *Cultural Studies 6*(1): 27–50.
Garber, Marjorie. 1993. *Vested Interests: Cross-Dressing and Cultural Anxiety*. New York and London: Routledge.
Garrard, Greg. 2004. *Ecocriticism*. London and New York: Routledge.
Gatens, Moira. 2000. "Feminism as 'Password'": Re-thinking the 'Possible' with Spinoza and Deleuze." *Hypatia 15*(2): 59–75.
Gelas, Bruno, and Hervé Micolet, eds. 2007. *Deleuze et les écrivians: littérature et philosophie*. Nantes: Ed. Cécile Default.
Gersdorf, Catrin, and Sylvia Mayer, eds. 2006. *Nature in Cultural and Literary Studies: Transatlantic Conversations on Ecocriticism*. Amsterdam: Rodopi.
Gijswijt-Hofstra, Marijke, ed. 1989. *Een schijn van verdraagzaamheid: afwijking en tolerantie in Nederland van de zestiende eeuw tot heden*. Hilversum: Verloren.
Ginkel, Rob van. 1999. *Op zoek naar eigenheid: denkbeelden en discussies over cultuur en identiteit in Nederland*. Den Haag: SdU Uitgevers.
Gleason, Henry Allen. 1926. "The Individualistic Concept of the Plant Association." *Bulletin of the Torrey Botanical Club* 53: 1–20.
Glissant, Edouard. 1981. *Le Discours Antillais*. Paris: Editions du Seuil.
———. 1989. *Caribbean Discourse: Selected Essays*. Translated and with an Introduction by J. Michael Dash. Charlottesville: University of Virginia Press.
———. 1990. *Poétique de la Relation*. Paris: Gallimard.
———. 1997. *Poetics of Relation*. Ann Arbor: University of Michigan Press.
Goodbody, Axel. 2006. "From Egocentrism to Ecocentrism: Nature and Morality in German Writing in the 1980s." In Catrin Gersdorf and Sylvia Mayer (eds.), *Nature in Cultural and Literary Studies: Transatlantic Conversations on Ecocriticism*, 393–414. Amsterdam: Rodopi.
Goodbody, Axel, and Catherine E. Rigby, eds. 2011. *Ecocritical Theory: New European Approaches*. Charlottesville: University of Virginia Press.
Goody, Jack. 1993. *The Culture of Flowers*. Cambridge: Cambridge University Press.
Gordon, David M., and Shepard Krech III. 2012. *Indigenous Knowledge and the Environment in Africa and North America*. Athens: Ohio University Press.
Goudzand Nahar, Henna. 2005. *Hele dagen in de regen* (Whole Days in the Rain). Breda: De Geus.
Gourdine, Angeletta K. M. 2003. *The Difference Place Makes: Gender, Sexuality and Diasporia Identity*. Columbus: Ohio State University Press.
Gowricharn, Rob. 1998. *Hollandse contrasten: Over de keerzijde van sociale integratie*. Leuven and Apeldoorn: Garant.
Graef, Ortwin de, and Henriette Louwerse. 2001. "De alteratie van Amsterdam: Hafid Bouazza's Liefdesgeschiedenissen." *Onverwerkt Europa: Jaarboek voor Literatuurwetenschap 1*: 169–91.
Grewal, Inderpal, and Caren Kaplan, eds. 1994. *Scattered Hegemonies: Transnational Feminist Practices*. Minneapolis: University of Minnesota Press.
Grosz, Elizabeth. 1993. "A Thousand Tiny Sexes: Feminism and Rhizomatics." *Topoi 12*: 167–79.
———. 1994. *Volatile Bodies: Toward a Corporeal Feminism*. Bloomington: Indiana University Press.
———. 2005. *Time Travels: Feminism, Nature, Power*. Durham, NC, and London: Duke University Press.
Halberstam, Judith, and Ira Livingston, eds. 1995. *Posthuman Bodies*. Bloomington and Indianapolis: Indiana University Press.

Halkes, Petra. 2001. *Aspiring to the Landscape*. Diss. Amsterdam: University of Amsterdam.
Hall, Stuart. 1980. "Race, Articulation, and Societies Structured In Dominance." In Marion O'Callaghan (ed.), *Sociological Theories: Race and Colonialism*, 305–45. Paris: UNESCO.
Hallward, Peter. 2001. *Absolutely Postcolonial: Writing Between the Singular and the Specific*. Manchester and New York: Manchester University Press.
Hansen, Miriam B. 1999. "Benjamin and Cinema: Not a One-Way Street." *Critical Inquiry 252*: 306–43.
Hardt, Michael, and Antonio Negri. 2000. *Empire*. Cambridge, MA: Harvard University Press.
Harré, Rom, Jens Brockmeier, Peter Mühlhäuser. 1999. *Greenspeak: A Study of Environmental Discourse*. Thousand Oaks, CA: Sage Publications.
Harries, Karsten. 1963. "Heidegger and Hölderlin: The Limits of Language." *The Personalist: A Quarterly Journal of Philosophy, Theology and Literature 44*(1): 5–23.
Harris, Wilson. 1983. *The Womb of Space: The Cross-Cultural Imagination*. Westport, CT: Greenwood Press.
Harvey, D. 1989. *The Condition of Postmodernity*. Oxford: Blackwell.
Hawley, John C., ed. 2001. *Postcolonial, Queer: Theoretical Intersections*. Albany: State University of New York Press.
Heidegger, Martin. 1971. *Poetry, Language, Thought*. Translated by Albert Hofstadter. New York: Harper & Row.
Heise, Ursula. 2008. *Sense of Place and Sense of Planet: The Environmental Imagination of the Global*. Oxford, New York: Oxford University Press.
Herbert, Christopher. 2004. "The Conumdrum of Coherence." *New Literary History 35*(2), 185–206.
Herman, Judith Lewis. 1994. *Trauma and Recovery: From Domestic Abuse to Political Terror*. London: Pandora.
Hondius, Dienke. 1999. *Gemengde huwelijken, gemengde gevoelens? Aanvaarding en ontwijking van etnisch en religieus verschil sinds 1945*. Den Haag: SdU Uitgevers.
Hopkinson, Nalo and Uppinder Mehan, eds. 2004. *So Long Been Dreaming: Postcolonial Science Fiction and Fantasy*. Vancouver, BC: Arsenal Pulp Press.
Hornung, A., and E. Ruhe, eds. 1998. *Postcolonialism and Autobiography: Michelle Cliff, David Dabydeen, Opal Palmer Adisa*. Amsterdam: Rodopi.
Hoving, Isabel. 2001a. *In Praise of New Travelers: Reading Caribbean Migrant Women's Writing*. Stanford, CA: Stanford University Press, 2001.
———. 2001b. "Hybridity: A Slippery Trail." In Joyce Goggin and Sonja Neef (eds.), *Travelling Concepts I: Text, Subjectivity, Hybridity*, 185–200. Amsterdam: ASCA Press.
———. 2002. "Remaining Where You Are: Kincaid and Glissant on Space and Knowledge." *Thamyris/Intersecting 9*(2002): 125–140.
———. 2003. "Corruptions, Mutations, Life: The Vegetable Imagination of Dutch Postcoloniality." In Anne Eggebert and Polly Gould (eds.), *Nature and Nation: Vaster than Empires*, 129–38. London: The London Institute/Central Saint Martins College of Art and Design, etc.
———. 2004. "Nat hout: Astrid Roemer's postkoloniale verbeelding" (Wet Wood: Astrid Roemer's Postcolonial Imagination). In Rosemarie Buikema and Maaike Meijer (eds.), *Kunsten in Beweging* (Arts in Movement), 323–41. Den Haag: SdU.
———. 2005. "Circumventing Openness: Creating New Senses of Dutchness." *Transit 1*(1): 1–11. http://german.berkeley.edu/transit/2005/TRANSIT50909.pdf
———. 2006. "On Invasions, Weeds and Wilderness: The Dutch Imagination of Globalisation (thrice)." In Clara A. B. Joseph and Janet Wilson (eds.), *Global Fissures: Postcolonial Futures*, 153–71. Amsterdam: Rodopi.
———. 2010. "On the Absence of Nature: Writing on Nature and Ecocriticism in the Netherlands." *Ecozon@: The European Journal of Literature, Culture and the Environment 1*(1): 167–73.
Hoving, Isabel, Hester Dibbits, and Marlou Schrover, eds. 2005. *Veranderingen van het Alledaagse*. Den Haag: SdU.
Huggan, Graham. 2004. "'Greening' Postcolonialism: Ecocritical Perspectives." *Modern Fiction Studies 50*(3): 701–33.

Huggan, Graham, and Helen Tiffin. 2010. *Postcolonial Ecocriticism: Literature, Animals, Environment*. London: Routledge.

Ingold, Tim. 2000. *The Perception of the Environment: Essays on Livelihood, Dwelling and Skill*. New York: Routledge.

JanMohamed, Abdul R. 1986. "The Economy of Manichean Allegory: The Function of Racial Difference in Colonialist Literature." In Henry Louis Gates, Jr. (ed.), *"Race", Writing, and Difference*, 78–106. Chicago: University of Chicago Press.

Jarman, Derek. 1995. *Derek Jarman's Garden*. New York: The Overlook Press.

Johnson, Erica L. 2009. "Ghostwriting Transnational Histories in Michelle Cliff's *Free Enterprise*." *Meridians: Feminism, Race, Transnationalism 9*(1): 114–39.

Jong, Mechtild Dominica Theresa Maria de. 2002. *Scheidslijnen in het denken over natuurbeheer in Nederland: een genealogie van vier ecologische theorieën*, Diss. Delft: University Press Science, 2002.

Joseph, Miranda. 2002. "Family Affairs: The Discourse of Global/Localization." In Arnaldo Cruz-Malavé and Martin F. Manalansan IV (eds.), *Queer Globalizations: Citizenship and the Afterlife of Colonialism*, 71–99. New York: New York University Press.

Kaiser, Birgit M., ed. 2014. *Singularity and Transnational Poetics*. New York and London: Routledge.

Kempadoo, Kamala. 2008. "Caribbean Sexuality: Mapping the Field." *OSO: Tijdschrift voor Surinamistiek en het Caribisch gebied 27*(1): 28–51.

Kempen, Michiel van, and Wim Rutgers. 2005.*Noordoostpassanten: 400 jaar Nederlandse Verhaalkunst over Suriname, de Nederlandse Antillen en Aruba* (Passers-By of the North-East Trade Wind: 400 Years Dutch Narrative Art on Surinam, the Dutch Antilles, and Aruba). Amsterdam and Antwerpen: Contact.

Kemperink, Mary. 2001. *Het verloren paradijs: de literatuur en de cultuur van het Nederlandse* fin de siècle. Amsterdam: Amsterdam University Press.

Keulartz, Jozef. 1998. *Struggle for Nature: A Critique of Radical Ecology*. London: Routledge.

Kincaid, Jamaica. 1984. *At the Bottom of the River*. London: Picador.

———. 1988. *A Small Place*. London: Virago.

———. 1996. *The Autobiography of My Mother*. New York: Farrar, Straus and Giroux.

———. 1997. *My Brother*. New York: Farrar, Straus and Giroux.

———. 1999. *My Garden (Book):*. New York: Farrar, Straus and Giroux.

———. 2002. *Mr. Potter*. New York: Farrar, Straus and Giroux.

———. 2005. *Among Flowers: A Walk in the Himalaya*. Washington, DC: National Geographic Society.

Kincaid, Jamaica, ed. 1998. *My Favorite Plant*. New York: Farrar, Straus and Giroux.

Kirkman, Robert. 2002. *Skeptical Environmentalism: The Limits of Philosophy and Science*. Bloomington: Indiana University Press.

Knutson, Lin. 1996. "Michelle Cliff and the Feminization of Space: 'No Telephone to Heaven' as Female Limbo Gateway into History." *Journal of Caribbean Studies 11*(3): 276–300.

Koenis, Sjaak. 2002. "De secularisatie en politisering van cultuur." In Jan Lucassen en Arie de Ruijter (eds.), *Nederland multicultureel en pluriform? Een aantal conceptuele studies*, 45–84. Amsterdam: Aksant.

Koopmans, Ruud. 2003. "Uitvluchten kan niet meer: een repliek op Böcker en Thränhardt." In *Migrantenstudies 1*: 45–56.

Kögler, Hans Herbert, and Karsten R. Stueber. 2000. *Empathy and Agency: the Problem of Understanding in the Human Sciences*. Boulder, CO: Westview Press.

Kolen, Jan and Ton Lemaire, eds. 1999. *Landschap in meervoud: Perspectieven op het Nederlandse landschap in de 20ste/21ste eeuw*. Utrecht: Jan van Arkel.

Kolodny, Annette. 1975. *The Lay of the Land*. Chapel Hill: University of North Carolina Press.

Kutzinski, Vera M. 2001."Improprieties: Feminism, Queerness, and Caribbean Literature." *Macalester International 11*: 165–206.

LaCapra, Dominick. 2001. *Writing History, Writing Trauma*. Baltimore, MD: Johns Hopkins University Press.

Lamb, Jonathan. 2001. "Modern Metamorphoses and Disgraceful Tales." *Critical Inquiry 28*: 133–66.

Lane, Christopher. 1996. "Uncertain Terms of Pleasure." *Modern Fiction Studies 42*(4): 807–26.

Lang-Peralta, Linda, ed. 2006. *Jamaica Kincaid and Caribbean Double Crossings.* Newark: University of Delaware Press.

Lauretis, Teresa de. 1984. *Alice Doesn't : Feminism, Semiotics, Cinema.* Bloomington: Indiana University Press.

Licops, Dominique. 2001. "Redefining Culture, Politicizing Nature: Negotiating the Essentialism of Natural Metaphors of Identification in the Work of James Clifford, Paul Gilroy and Aimé Césaire." *Thamyris/Intersecting 8*: 53–68.

Lionnet, Françoise. 1992. "Of Mangoes and Maroons: Language, History, and the Multicultural Subject of Michelle Cliff's *Abeng*." In Sidonie Smith and J. Watson (eds.) *De/Colonizing the Subject: The Politics of Gender in Women's Autobiography*, 321–47. Minneapolis: University of Minnesota Press.

Lohman, Larry. 1993. "Resisting Green Globalism." In Wolfgang Sachs (ed.), *Global Ecology: A New Arena of Political Conflict*, 157–69. London: Zed Books.

Loomba, Ania. 2005. *Colonialism/Postcolonialism.* Abingdon, Oxon; New York: Routledge.

Luckhurst, Roger. 2008. *The Trauma Question*. Abingdon and New York: Routledge.

Lyotard, Jean-François. 1992. "Answering the Question: What Is the Postmodern?" *The Postmodern Explained to Children*. Sydney: Power Publications. http://www.marginalutility.org/wp-content/uploads/2010/04/machete_reading_may1_c.pdf. Accessed 28 April 2015.

MacDonald-Smythe, A. 2001. *Making Homes in the West/Indies: Constructions of Subjectivity in the Writings of Michelle Cliff and Jamaica Kincaid*, New York: Garland.

Macnaghten, Phil, and John Urry. 1998. *Contested Natures*. London and Thousand Oaks, CA: SAGE Publications.

Madou, Jean-Pol. 1996. *Edouard Glissant: De mémoire d'arbres.* Amsterdam and Atlanta: Rodopi.

Maes-Jelinek, Hena. 2005. "From Living Nature to Borderless Culture in Wilson Harris's Work." In Elizabeth M. DeLoughrey, Renée K. Gosson, and George B. Handley (eds.), *Caribbean Literature and the Environment: Between Nature and Culture*, 247–260. Charlottesville: University of Virginia Press.

Maisonnat, Claude. 2001. "The Poetics of Death: Olive Senior's 'Tears of the Sea.'" In Jacqueline Bardolph (ed.), *Telling Stories: Postcolonial Short Fiction in English*, 119–28. Amsterdam: Rodopi.

Malpas, Simon, and Paul Wake. 2006. *Routledge Companion to Critical Theory.* Abingdon, Oxon, and New York: Routledge.

Marks, Laura U. 2000. *The Skin of the Film: Intercultural Cinema, Embodiment, and the Senses*. Durham, NC: Duke University Press.

Marshall, Alan. 2002. *The Unity of Nature: Wholeness and Disintegration in Ecology and Science*. London: Imperial College Press.

Massey, Doreen. 1994. *Space, Place, and Gender*. Minneapolis: University of Minnesota Press.

McClintock, Anne. 1995. *Imperial Leather: Race, Gender, and Sexuality in the Colonial Contest.* New York, London: Routledge.

McLeod, John. 2000. *Beginning Postcolonialism.* Manchester: Manchester University Press.

McWhorter, Ladelle. 2010. "Enemy of the Species." In Mortimer-Sandilands and Erickson, 73–101.

Melas, Natalie. 2001. "Re-imagining the Universal." In Salah Hassan and Iftikhar Dadi (eds.), *Unpacking Europe: Towards a Critical Reading*, 134–51. Rotterdam: Museum Boijmans Van Beuningen/NAi Publishers.

———. 2007. *All the Difference in the World: Postcoloniality and the Ends of Comparison.* Stanford, CA: Stanford University Press.

Menocal, Maria Rosa. 1987. *The Arabic Role in Medieval Literary History: A Forgotten Heritage.* Philadelphia: University of Pennsylvania Press.

Mercer, Kobena. 2002. "Skin Head Sex Thing: Racial Difference and the Homoerotic Imaginary." In Rachel Adams and David Savran (eds.), *The Masculinity Studies Reader*, 188–200. Malden, MA: Wiley-Blackwell.

Merchant, Carolyn. 1983. *The Death of Nature: Women, Ecology, and the Scientific Revolution* (1980). San Francisco: Harper and Row.
———. 1995. *Earthcare: Women and the Environment*. New York: Routledge.
Meulen, Dik van der. 2009. *Het bedwongen bos: Nederlanders and hun natuur*. Amsterdam: SUN.
Miller, Daniel. 1991. *The Myth of Primitivism: Perspectives on Art*. Ed. Susan Hiller. London, New York: Routledge.
Miller, Jacques-Alain. 2001. "Lacanian Biology and the Event of the Body." *Lacanian Ink 18*: 6–29.
Milton, Kay. 1993. *Environmentalism: The View from Anthropology*. London, New York: Routledge.
Mitchell, W. J. T. 1994. *Landscape and Power*. Chicago and London: University of Chicago Press.
Mitsch, Ruthmarie H. 1997. "Maryse Condé's Mangroves." *Research in African Literatures 28*(4): 54–70.
Moor, Els. 2004. "De ondraaglijke zwaarte van het bestaan. Het proza van zeven Surinaamse schrijfsters uit de 21ste eeuw" (The Unbearable Heaviness of Being. Prose by Seven Surinamese Writers in the 21st Century). *OSO: Tijdschrift voor Surinamistiek 32*(1): 110–21.
Mootoo, Shani. 1996. *Cereus Blooms at Night*. Vancouver, BC: Gang Press Publishers.
Mordecai, Pamela, and Betty Wilson, eds. 1990. *Her True-True Name*. Oxford: Heinemann.
Mortimer-Sandilands, Catriona. 2005. "Unnatural Passions? Notes Toward a Queer Ecology." *Invisible Culture: An Electronic Journal for Visual Culture 9*: 1–21. https://www.rochester.edu/in_visible_culture/Issue_9/issue9_sandilands.pdf
———. 2008. "Queering Ecocultural Studies." *Cultural Studies 22*(1): 455–76.
Mortimer-Sandilands, Catriona, and Bruce Erickson, ed. 2010. *Queer Ecologie: Sex, Nature, Politics, Desire*. Bloomington and Indianapolis: Indiana University Press.
Morton, Timothy. 2007. *Ecology without Nature: Rethinking Environmental Aesthetics*. Cambridge, MA: Harvard University Press.
Murphy, Patrick D. 2012. "An Ecological Feminist Revisioning of the Masculine Sublime." *Revista Canaria de Estudios Ingleses 64*: 79–94.
Narain, Denise deCaires. 1996. "Olive Senior." In Bernth Lindfors (ed.), *Twentieth-Century Caribbean and Black African Writers, Third Series*, 340–48. Detroit: Gale Press.
Nelson, Deborah. 2006. "The Virtue of Heartlessness: Mary McCarthy, Hannah Arendt, and the Anesthetics of Empathy." *American Literary History 18*(1): 86–101.
Nixon, Rob. 2005. "Environmentalism and Postcolonialism." In Ania Loomba et al. (eds.), *Postcolonial Studies and Beyond*, 233–51. Durham, NC: Duke University Press.
North, Michael. 1995. "Modernism's African Mask: The Stein-Picasso Collaboration." In Elazar Barkan and Ronald Bush (eds.), *Prehistories of the Future: The Primitivist Project and the Culture of Modernism*, 270–89. Stanford, CA: Stanford University Press.
O'Brien, Susie. 2001. "Articulating a World of Difference: Ecocriticism, Postcolonialism and Globalization." *Canadian Literature 170–171*: 140–58.
———. 2002. "The Garden and the World: Jamaica Kincaid and the Cultural Borders of Ecocriticism." *Mosaic: A Journal for the Comparative Study of Literature and Ideas 35*(2): 167–84.
O'Callaghan, Evelyn. 1988. "Feminist Consciousness: European/American Theory, Jamaican Stories." *Journal of Caribbean Studies 6*(2), 143–62.
Ombre, Ellen. 1994. *Vrouwvreemd*. Amsterdam: Arbeiderspers.
———. 2000. *Valse verlangens*. Amsterdam: Arbeiderspers.
Paquet, Sandra Pouchet. 2002. *Caribbean Autobiography: Cultural Identity and Self-Representation*. Madison: University of Wisconsin Press.
Parekh, Bhikhu. 2003. "Wat is multiculturalisme"? *Eutopia 4*: 7–14.
Patteson, Richard F. 1993. "The Fiction of Olive Senior: Traditional Society and the Wider World." *ARIEL: A Review of International English Literature 24*(1), 13–33.
Pecic, Zoran. 2013. *Queer Narratives of the Caribbean Diaspora: Exploring Tactics*. Houndmills: Palgrave MacMillan.

Philip, Marlene Nourbese. 1997. *A Genealogy of Resistance and Other Essays.* Toronto: Mercury Press.
Phillips, Dana. 1999. "Ecocriticism, Literary Theory, and the Truth of Ecology." *New Literary History 30*(3): 577–602.
Phillips, Richard, Diane West and David Shuttleton, eds. 2000. *De-Centering Sexualities: Politics and Representation Beyond the Metropolis*. London: Routledge.
Plumwood, Val. 1993. *Feminism and the Mastery of Nature.* London: Routledge.
———. 2007. "Journey to the Heart of Stone." In Fiona Beckett and Terry Gifford (eds.), *Culture, Creativity, and Environment: New Environmentalist Criticism*, 17–36. Amsterdam, New York: Rodopi.
Pollard, Velma. 1988. "An Introduction to the Poetry and Fiction of Olive Senior." *Callaloo 11*(3): 540–46.
———. 1991. "Mothertongue Voices in the Writing of Olive Senior and Lorna Goodison." In S. Nasta (ed.), *Motherlands*, 238–53. London: Women's Press.
Pratt, Mary Louise. 1992. *Imperial Eyes: Travel Writing and Transculturation.* London: Routledge.
Prins, Baukje. 2004. *Voorbij de onschuld: het debat over integratie in Nederland.* Amsterdam: Van Gennep.
Pulsipher, Lydia Mihelic. 1994. "The Landscapes and Ideational Roles of Caribbean Slave Gardens." In Naomi F. Miller and Kathryn L. Gleason (eds.), *The Archaeology of Garden and Field*, 202–21. Philadelphia: University of Pennsylvania Press.
Purdom, Judy. 1995. "Mapping Difference." *Third Text 32*: 19–32.
Rahim, Jennifer. 2005. "(Not) Knowing the Difference: Calypso Overseas and the Sound of Belonging in Selected Narratives of Migration." *Anthurium: A Caribbean Studies Journal 3*(2). http://scholar.library.miami.edu/anthurium/volume_3/issue_2/rahim-notknowing.htm
Reiss, Timothy. 2002. *Against Autonomy: Global Dialectics of Cultural Exchange.* Stanford, CA: Stanford University Press.
Richards, Constance S. 2005. "Nationalism and the Development of Identity in Postcolonial Writing: Zoë Wicomb and Michelle Cliff." *Research in African Literatures 36*(1): 20–33.
Rigby, Kate. 2004. "Earth, World, Text: On the (Im)possibility of Ecopoiesis." *New Literary History 35*: 427–42.
Roberston, Roland. 1992. *Globalization: Social Theory and Global Culture*. London: Sage.
Roemer, Astrid. 1997. *Lijken op liefde*. Amsterdam: Arbeiderspers.
———. 1998. *Was Getekend.* Amsterdam: Arbeiderspers.
———. 2001. *Roemers drieling: Romantrilogie*. Amsterdam: Arbeiderspers.
Rohlehr, Gordon. 1981. *Pathfinder: Black Awakening in* The Arrivants *of Edward Kamau Brathwaite*. Port of Spain: College Press.
Roos, Bonnie, and Alex Hunt, eds. 2010. *Postcolonial Green: Environmental Politics and World Narratives*. Charlottesville: University of Virginia Press.
Rosendale, Steven, ed. 2002. *The Greening of Literary Scholarship: Literature, Theory, and the Environment*. Iowa City: University of Iowa Press.
Roughgarden, Joan. 2004. *Evolution's Rainbow: Diversity, Gender, and Sexuality in Nature and People*. Berkeley: University of California Press.
Ruffin, Kimberley N. 2010. *Black on Earth: African American Ecoliterary Traditions*. Athens: University of Georgia Press.
Russell, Sharman Apt. 2001. *Anatomy of a Rose: the Secret Life of Flowers*. London: Heinemann.
Sachs, Wolfgang, ed. 1993. *Global Ecology: A New Arena of Political Conflict*. London: Zed Books.
Santos Febres, Mayra. 2004. "Caribe y travestismo." In Rita De Maeseneer and An Van Hecke (eds.), *El artista caribeño como guerrero de lo imaginario*, 37–44. Madrid: Iberoamericana.
Sarduy, Severo. 1989. *Written on a Body*, translated by Carol Maier. New York: Lumen Books.
Schama, Simon. 1995. *Landscape and Memory*. New York: Knopf.
Scheffer, Paul. 2000. "Het multiculturele drama: een repliek." *NRC Handelsblad*, January 29, 2000.

Scheler, Max. 1992. *On Feeling, Knowing, and Valuing*. Edited by Harold J. Bershady. Chicago and London: University of Chicago Press.

Schiebinger, Londa. 1993. *Nature's Body: Gender in the Making of Modern Science.* Boston: Beacon Press.

Schoo, H. J. 2000. *De verwarde natie: dwarse notities over immigratie in Nederland.* Amsterdam: Prometheus.

Schouten, Matthijs G. C. 2001. *De natuur als beeld in religie, filosofie en kunst.* Utrecht: KNNV Uitgeverij.

Schroevers, Piet. 1999. "Menselijk bemiddelde natuur." In Jan Kolen and Ton Lemaire (eds.), *Landschap in meervoud: Perspectieven op het Nederlandse landschap in de 20ste/21ste eeuw*, 141–60. Utrecht: Jan van Arkel.

Schwenger, Peter. 2001. "Words and the Murder of the Thing." *Critical Inquiry 28*: 99–113.

Scott, Lawrence. 1992. *Witchbroom*. London: Allison and Busby.

———. 1998. *Aelred's Sin*. London: Allison and Busby.

———. 2013. "Tales Told Under the San Fernando Hill." In Fumagalli, Ledent, and del Valle Alcalá (eds.), *The Cross-Dressed Caribbean: Writing, Politics, Sexualities*, 186–200. Charlottesville: University of Virginia Press.

Senior, Olive. 1985. *Talking of Trees*. Jamaica: Calabash.

———. 2005. *Gardening in the Tropics*. Toronto: Insomniac Press.

Seymour, Nicole. 2013. *Strange Natures: Futurity, Empathy, and the Queer Ecological Imagination*. Urbana, Chicago, and Springfield: University of Illinois Press.

Shepherd, Verene A. and Glen L. Richards, eds. 2002. *Questioning Creole: Creolisation in Discourses in Caribbean Culture*. Kingston: Ian Randle Publishers.

Shuttleton, David. 2000. "The Queer Politics of Gay Pastoral." In Richard Phillips, Diane West, and David Shuttleton (eds.), *De-Centering Sexualities: Politics and Representation beyond the Metropolis*, 125–46. London: Routledge.

Shewry, Teresa. 2014. "Geologic Fictions: Science, Imperialism, and Transnational Literatures of Deep Time." *ISLE: Interdisciplinary Studies in Literature and Environment 21*(2), 253–70.

Shiva, Vandana. 1993. "The Greening of the Global Reach." In Wolfgang Sachs (ed.), *Global Ecology: A New Arena of Political Conflict*, 149–57. London: Zed Books.

Shohat, Ella, and Robert Stam. 1994. *Unthinking Eurocentrism: Multiculturalism and the Media.* London and New York: Routledge.

Silverman, Kaja. 1983. *The Subject of Semiotics*. New York and Oxford: Oxford University Press.

Singh, Vandana. 2004. "Delhi." In Nalo Hopkinson and Uppinder Mehan (eds.), *So Long Been Dreaming: Postcolonial Science Fiction and Fantasy*, 79–94. Vancouver: Arsenal Pulp Press.

Slobbe, Thomas van. 2004. *Dagboek van een lege plek*. Beek-Ubbergen: Uitgeverij wAarde.

Smith, Derik, and Cliff Beumel. 2006. "My Other: Imperialism and Subjectivity in Jamaica Kincaid's *My Brother*." In Linda Lang-Peralta (ed.), *Jamaica Kincaid and Caribbean Double Crossings*, 96–112. Newark: University of Delaware Press.

Smith, Jennifer J. 2009. "Birthed and Buried: Matrilineal History in Michelle Cliff's *No Telephone to Heaven*." *Meridians: Feminism, Race, Transnationalism 9*(1): 141–62.

Smith, Sidonie, and J. Watson, eds. 1992. *De/Colonizing the Subject: The Politics of Gender in Women's Autobiography*. Minneapolis: University of Minnesota Press.

Smyth, Gerry, and Jo Croft. 2006. *Our House: the Representation of Domestic Space in Modern Culture*. Amsterdam and New York: Rodopi.

Smyth, Heather. 1999. "Sexual Citizenship and Caribbean-Canadian Fiction: Dionne Brand's 'In Another Place, Not Here' and Shani Mootoo's *Cereus Blooms at Night*." *ARIEL: A Review of International English Literature 30*(2): 141–60.

Soja, Edward. 1996. *Thirdspace: Journeys to Los Angeles and Other Real-and-Imagined Spaces*. Cambridge, MA: Blackwell.

Spillers, Hortense J. 1987. "Mama's Baby, Papa's Maybe: An American Grammar Book." *Diacritics 17*, 65–81.

Spivak, Gayatri Chakravorty. 1993. *Outside in the Teaching Machine.* New York: Routledge.

———. 1999. *A Critique of Postcolonial Reason: Toward a History of the Vanishing Present*. Cambridge, MA, and London: Harvard University Press.

———. 2003. *Death of a Discipline*. New York: Columbia University Press.

Spurr, David. 1993. *The Rhetoric of Empire: Colonial Discourse in Journalism, Travel Writing, and Imperial Administration*. Durham, NC: Duke University Press.

Steinfeld, J. I. 1944. Phänomenologisches Und Psychoanalytisches Zum Problem Des Mitleids. (Phenomenology and Psychoanalysis in the Problem of Sympathy). *Psychoanalytical Quarterly 13*: 120–121.

Stepan, Nancy. 1982. *The Idea of Race in Science: Great Britain 1800–1960*. London: Macmillan.

Stoler, Ann Laura. 1995. *Race and the Education of Desire: Foucault's History of Sexuality and the Colonial Order of Things*. Durham, NC: Duke University Press.

———. 2002. *Carnal Knowledge and Imperial Power: Race and the Intimate in Colonial Rule*. Berkeley: University of California Press.

Stouck, Jordan. 2005. "Gardening in the Diaspora: Place and Identity in Olive Senior's Poetry." *Mosaic 38*(4): 103–22.

Strongman, Roberto. 2007. "Postmodern Developments in Michelle Cliff's 'No Telephone to Heaven' and Esmeralda Santiago's 'When I was Puerto Rican.'" *Journal of Caribbean Studies 21*(3): 189–98.

Stryker, Susan, and Aren Z. Aizura, eds. 2013. *The Transgender Studies Reader 2*. New York: Routledge.

Stryker, Susan and Stephen Whittle, eds. 2006. *The Transgender Studies Reader*. New York: Routledge.

Suurmond, Jeanine. 1998. *Grenzen aan grenzen: een analyse van politieke debatten over vluchtelingen en migranten in termen van zorg en rechtvaardigheid.* Maastricht: Shaker.

Swaan, Abram de. 2006. *De botsing der beschavingen en de strijd der geslachten.* Utrecht: Forum en Ger Guijs.

Taussig, Michael. 1993. *Mimesis and Alterity: A Particular History of the Senses*. Routledge: New York.

Thomas, Greg. 2007. *The Sexual Demon of Colonial Power: Pan-African Embodiment and Erotic Schemes of Empire.* Bloomington and Indianapolis: Indiana University Press.

Tinsley, Natasha. 2005. "Rosebud Is My Mama, Stanfaste Is My Papa: Hybrid Landscapes and Sexualities in Surinamese Oral Literature." In Elizabeth M. DeLoughrey, Renée K. Gosson, and George B. Handley (eds.), *Caribbean Literature and the Environment: Between Nature and Culture*, 169–81. Charlottesville: University of Virginia Press.

Tompkins, Peter, and Christopher Bird. 1974. *The Secret Life of Plants*. Harmondsworth: Penguin.

Toorn, Willem van. 2004. "Waar wij wonen." In van Eeden (ed.), *Want je ziet niet alleen wat je ziet: Over het werk van Koos van Zomeren*, 39–54. Amsterdam: Arbeidspers.

———. 2008. "Landschap als geheugen." In Joost Zwagerman (ed.), *De Nederlandse en Vlaamse literatuur vanaf 1880 in 200 essays*, 690–97. Amsterdam: Prometheus.

Tuana, Nancy. 2008. "Viscous Porosity: Witnessing Katrina." In Stacy Alaimo and Susan Hekman (eds.), *Material Feminisms*, 188–213. Bloomington: Indiana University Press.

Vakoch, Douglas A. 2012. *Feminist Ecocriticism: Environment, Women, and Literature*. Lanham, MD: Lexington Books.

Vasey, Paul, and Volker Sommer, eds. 2006. *Homosexuality in Animals: An Evolutionary Perspective*. Cambridge: Cambridge University Press.

Verstraete, Ginette. 2002. "Heading for Europe: Tourism and the Global Itinerary of an Idea." *Thamyris/Intersecting 9*: 33–52.

Vertovec, Steven, and Robin Cohen, eds. 1999. *Migration, Diaspora's and Transnationalism*. Cheltenham: Edward Elgar.

Visser, Irene. 2011. "Trauma Theory and Postcolonial Studies." *Journal of Postcolonial Writing 47*(3): 270–82.

Vries, Annette de. 2002. *Scheurbuik* (Scurvy). Amsterdam and Antwerpen: Atlas.

Vuijsje, Herman. 1986. *Vermoorde onschuld: Etnisch verschil als Hollands taboe*. Amsterdam: Bert Bakker.

Waarover we niet moeten praten: nieuwe Surinaamse en Antilliaanse verhalen (What We Should Not Talk about: New Surinamese and Dutch-Antillian Stories). 2007. Haarlem: In de Knipscheer.

Walker, Julian. 2003. "The Language of Encroachment." In Anne Eggebert and Polly Gould (eds.), *Nature and Nation: Vaster Than Empires*, 115–22. London: The London Institute/ Central Saint Martins College of Art and Design, etc.

Walters, Wendy. 1998. "Michelle Cliff's *No Telephone to Heaven*: Diasporic Displacement and the Feminization of the Landscape." In Elazar Barkan and Marie Denise Shelton (eds.), *Borders, Exiles, Diasporas*, 217–33. Stanford, CA: Stanford University Press.

Warren, Karen J., ed. 1994. *Ecological Feminism*. London and New York: Routledge.

Watts, David. 1987. *The West Indies: Patterns of Development, Culture and Environmental Change since 1492.* Cambridge: Cambridge University Press.

Weizäcker, Christine von. 1993. "Competing Notions of Biodiversity." In Wolfgang Sachs (ed.), *Global Ecology: A New Arena of Political Conflict*, 117–31. London: Zed Books.

Wekker, Gloria. 1994. *Ik ben een gouden munt, ik ga door vele handen, maar verlies mijn waarde niet. Subjectiviteit en seksualiteit van Creoolse volksklasse vrouwen in Paramaribo.* Amsterdam: Vita.

———. 2006. *The Politics of Passion: Women's Sexual Culture in the Afro-Surinamese Diaspora.* New York: Columbia University Press.

Werbner, Pnina, and Tariq Modood, eds. 1997. *Debating Cultural Hybridity: Multi-Cultural Identity and the Politics of Anti-Racism*. London: Zed Books.

Westling, Louise. 2006. "Literature, the Environment, and the Question of the Posthuman." In Catrin Gersdorf and Sylvia Mayer (eds.), *Nature in Cultural and Literary Studies: Transatlantic Conversations on Ecocriticism*, 25–48. Amsterdam: Rodopi.

Wilson, Edward O. 1998. *Consilience: The Unity of Knowledge*. New York: Knopf.

Wouters, Cas. 1990. *Van minnen en sterven: informalisering van omgangsvormen rond seks en dood.* Diss. UVA. Amsterdam.

Wroe, Emily. 2004. "'A Strange Synchronicity': The Language of Landscapes in Lawrence Scott's *Aelred's Sin*." In Sandra Courtman (ed.), *Beyond the Blood, the Beach and the Banana: New Perspectives in Caribbean Studies*, 426–43. Kingston: Ian Randle.

Young, Robert. 1995. *Colonial Desire: Hybridity in Theory, Culture and Race.* London: Routledge.

Ziarek, Krzystof. 2002. "Is All Technological? Global Power and Aesthetic Forces." *Centennial Review 2*(3), 139–68.

Žižek, Slavoj. 1992. *Looking Awry: An Introduction to Jacques Lacan through Popular Culture.* Cambridge, MA: MIT Press.

———. 2001. "The Rhetorics of Power." *Diacritics 31*(1): 91–104.

Index

About the Author

Isabel Hoving is chief diversity officer at Leiden University. She is also associate professor at the Department of Film and Literary Studies and the Leiden University Centre for the Arts in Society, where she specializes in gender studies, interculturality, the environmental humanities, and games studies. Her publications include a study on Caribbean migrant women writers, *In Praise of New Travelers* (Stanford University Press, 2001), coedited books on (Dutch) migration, Caribbean literatures, African literature and art; an edited volume, *Dutch Racism*, with Philomena Essed (Rodopi, 2014), and (as coauthor) a book on Dutch literary writing on multiculturality (ACCO, 2015). She is member of the editorial boards of Thamyris/ Intersecting: Place, Sex, and Race; Ecocritical Theory and Practice, and the advisory boards of Ecozon@: The European Journal on Literature, Culture and the Environment; and Studien zu Literatur, Kultur und Umwelt/Studies in Literature, Culture, and the Environment. In 2012, she launched the Benelux Association for the Study of Literature, Culture, and Environment, BASCE. In addition to her academic work, she is an awarded writer of crossover and adult philosophical fantasy.

www.ingramcontent.com/pod-product-compliance
Ingram Content Group UK Ltd.
Pitfield, Milton Keynes, MK11 3LW, UK
UKHW041632031026
15587UKWH00001B/228

9 781498 526753